Introduction to Air Navigation

A technical and operational approach

Javier Lloret

Air Traffic Controller. Aeronautical Engineer

Enaire. Universidad Carlos III de Madrid

Javier Lloret [Ed]

Printed by Create Space

Madrid, June 2017

©Javier Lloret
First edition August 2015
Second edition January 2016
Third edition June 2017
Printed by Create Space

A mi padre, Pepe,

por transmitirme su pasión por el vuelo

Contents

Chapter 11. Safety   143

Chapter 12. Cartography   155

Chapter 13. Area Navigation   175

Bibliography   229

Chapter 1. History and basics

1 Introduction

Along the following chapters we will be covering different topics related to one common particular field of aerospace engineering: Air Navigation. Air Navigation is the science that includes all the necessary steps to conduct/move an aircraft from one initial point to another final point in a safe and efficient manner. From the point of view of each individual aircraft, there are two basic questions we have to address:

- Where am I? - we won´t be able to navigate if we can´t determine our position anytime. Determination and understanding of current position is named *position awareness*[1].

- What shall I do to fly towards my destination (safely and efficiently)? - we will also have to navigate, choose the way we want to steer our aircraft, following a determined route in order to fulfill certain requirements of safety and efficiency.

General navigation is also performed in other transport modes (e.g. sailing a ship), but air navigation is developed within a particular mean (i.e. air) and in an environment (boundary conditions) that makes the system face particular issues that have to be tackled:

- There is no need for ground infrastructure while airborne. This basically means that aircraft have a lot more freedom of movement than other transport vehicles (e.g. cars or trains that have to follow determined paths on earth, given by roads or railways).

---

[1] Even though it may sound simple, getting lost is a lot more common than it may seem (specially among general aviation pilots). Loss of position awareness may lead to fatal consequences, specially due to endurance restrictions associated to the limited fuel onboard

- Fuel imposes important limitations to flight duration. Range and endurance of aircraft are limited. For this reason, lack of proper position awareness is of extreme danger, given the fact that running out of fuel may lead to forced gliding onto non-prepared ground fields (in many cases with fatal consequences).

- Aircraft need speed to fly and can't stop in the middle of the air. Stopping aside to think or ask someone about directions is something we can do when we are in our cars, but never an option when flying an aircraft. When onboard, pilots rely on instruments and any external aid they may get will come from sources such as navaids, radio-communications, visual observance or radars (that can detect aircraft's position).

- The mean through which the aircraft moves (air) strongly affects the trajectory and performance of the vehicle. Presence of weather phenomena will affect the chosen trajectory for the sake of efficiency (e.g. wind profiles change everyday, making some trajectories more efficient than others) and safety (e.g. presence of turbulent clouds such as CBs, ice, hail, fog or even volcanic ashes).

- Aircraft move faster than other transport vehicles. This makes times shorter and reduces the time pilots and air traffic controllers have to think before making decisions. Understanding a situation properly is named *situation awareness*. The faster we have to make our decisions, and the shorter the times we have to think about these decisions, the higher the likelihood to lose our situational awareness. In aviation particularly, there is very small room for omissions or forgetfulnesses, since they can have very fatal consequences.

Besides these five differences we find between air navigation and general navigation (related to other transport modes), we can find one important similarity:

- Danger of hitting other vehicles or obstacles. Similarly to ground or sea transportation, aircraft may find obstacles to avoid when moving. It is because of this that air navigation has to tackle the issue of avoiding aircraft collisions (between each other and between aircraft and terrain or obstacles). We will get deeper into this along the following chapters, describing the implemented services and techniques that are used to avoid collisions. The main two events that are highly likely to down an aircraft are onboard fire and aircraft collision (resulting in structural damage). The loss of position awareness in a context of poor visibility (presence of clouds, fog or night time), as we will see through the text, may lead to aircraft unintentionally flying towards an obstacle (e.g. mountain), this kind of events, when they have fatal consequences, are named Controlled Flight Into Terrain events (CFITs).

1.1   History of Air Navigation

Not long after the Wright brothers took off for the first time in North Carolina did men start wondering about navigation issues. By the beginning of the 20th century, mankind had already developed complex and precise techniques that drove maritime navigation, and that were to be applied also (when possible) in air navigation. At the beginning traffic levels were so low that main concerns focused on having enough position awareness and a correct aircraft guidance. It was just navigation for the individual, not a complex network as we know it today.

First flights navigated under complete visual references, hardly supported by a map and a compass. Visual navigation requires good meteorological conditions as well as daylight, which imposed restrictions to air transport. Looking for one village, a certain river, railways or any other significant orographic element was the day to day of pilot´s navigation concerns. Needless is to say (for those readers that are not familiar with visual navigation) that from an aircraft it is not easy to decipher what there is below, specially when the pilot is not familiarized with the area[2]. And even when position awareness is successfully achieved, visual navigation normally requires flying longer routes (depending on where the references that are expected to be identified are placed on Earth), lowering down flight efficiency, increasing fuel consumption and flight times. As we will see in section 1.2, dead reckoning techinques were used to fly shorter routes or over areas where no references were available.

For longer flights, when no references were available, different maritime techniques were used. In the maritime world, astrolabes or sextants were used to determine latitude (measuring and checking tables for the angles at which different celestial bodies were observed), and tools such as accurate clocks were used to estimate longitude (changing time of noon indicates how many meridians a boat moves per day).

After WWI, light beacons (similar to lighthouses for boats) were used enabling night flights as well as torches and fire places to show the way. Radio communication equipments were finally installed onboard when state of the art radio weights met aircraft requirements, enabling surveillance beyond observation from the ground (through position radio reports made by pilots) and radiogoniometry. Radiogoniometry lets stations on the ground find out the direction from which they are receiving a radio transmission.

Improvements in technology allowed the implementation of new ground-based references that could help aircraft navigate, basically through the use of electromagnetic fields that instruments could receive and treat, deciphering information about their relative position to these ground-based navigation aids (commonly named *navaids*).

One of the first implemented navaids was the Low Frequency Radiorange (LFR), which was extensively used by aircraft in the 30s and 40s. Pairs of ground

---

[2]We could here recall that there are two types of pilots, those who have already gotten lost, and those who will

stations emitted morse signals (that pilots had to listen to, see Figure [1.1]) which were coded differently for given directions, resulting overlapped when flying the straight line between two stations thus letting pilots know whether they had to turn right or left until being centered on the line that joined the two stations. The appearance and deployment of VORs from the 40s on, meant that by 1970 most LFRs had disappeared, remaining none of them in use today.

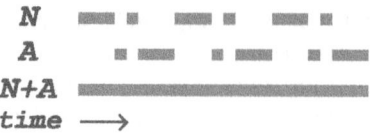

Figure 1.1: LFR Audio Signals © Crum375 / Wikimedia Commons / CC BY-SA 3.0

NDB, VOR, DME and ILS are navaids that have been broadly implemented, supporting air navigation during the last 50 years, enhancing safety and efficiency of instrumental flights, taking air transport to the levels of development and reliability we know today. Area navigation (RNAV) started appearing in the 70s, enabling aircraft fly shorter routes, not being forced to fly from one navaid to another. Current deployment of new technologies, enabling GNSS (Global Navigation Satellite System), strongly threatens the survival of old-fashioned ways to understand air navigation. Programs such as SESAR (in Europe) or NextGen (in US) are expected move air navigation towards new paradigms where satellite-based navigation will play a key role.

However, air navigation today does not only tackle problems regarding positioning and steering of one individual aircraft, it also deals with the performance of a whole interacting network (set of flights that want to use the same airspace at the same time), providing solutions. At the beginning of aviation, as we have seen, not getting lost and avoiding terrain obstacles was already a big challenge, very low levels of traffic imposed no concerns about finding other aircraft on the way (number of conflicts was extremely low). But little by little traffic started increasing, specially in the vicinity of aerodromes (at first), forcing the initial navigation concept to embrace the concept of circulation (organization of flux). Basic rules were defined at first, based on the principle of being able to see and be seen. First rules of the air made pilots responsible for all traffic separations, but as traffic levels kept on increasing, the role of an air traffic controller became necessary. Tower air traffic controllers appeared first to preserve safety, order and efficiency in the vicinity of aerodromes, later appearing the first area control centers (USA, 1935) to solve enroute conflicts.

As a result of this, new legal frameworks became necessary to regulate air navigation, developing the different technical, operational and juridic aspects. The

Figure 1.2: First tower and area air traffic controllers © Avstop

ATM-CNS concept, that we will cover along the following chapters, was developed.

1.2 The dead reckoning technique

The most simple technique for navigation consists of extrapolating aircraft's position along time given an initial known position and flight values of speed and time. From very simple kinematics we know that $v = \frac{\Delta s}{\Delta t}$, so knowing our speed ($v$) and measuring time with a chronometer (Δt), maintaining a determined heading (rhumb), we can estimate our future position (computing Δs). This very simple technique lets us fly different legs along which there is no visual reference nor position awareness. Everytime the pilot recognizes he has overflown a visual checkpoint, he resets the chronometer, changes the heading and faces towards the next target.

This technique becomes less and less reliable the longer the length of the legs because of the following sources of error:

- Instrument errors

    - Anemometer. As we will see in following chapters, aircraft anemometers measure Indicated Airspeed, which is different from True Air Speed (the one that the aircraft truly has with respect to the mass of air around it). The difference between both increases the higher the altitude of the aircraft (the less dense air becomes).
    - Compass. Not only should the pilot care about magnetic deviation as we will see in section 2, but also about any internal magnetic fields created inside the cockpit (by some other electronic instruments) that affect the indication of the compass. Different accelerations of the vehicle (linear and angular) do also distort compass indications.
    - Clock. Errors associated to time measurement also affect position estimation.

- Wind deviations. Maintaining a constant heading in the presence of lateral wind will make the aircraft drift, not reaching the target. We will get deeper into this in section 3. Also, tail or headwind components will affect the time it takes to reach the target. Wind predictions for different vertical levels are part of meteorological forecasts nowadays, but they were not at the beginning of aviation. Also, current Inertial Navigation Systems onboard (assisted by satellite-based tools for the determination of position) help solve the equation posed in section 3, giving information about actual wind intensity and direction at the position of the aircraft.

- Pilot errors. Human errors associated to maintaining constant speed and heading introduce uncertainty into the navigation process.

Because of the enormous uncertainty associated to this technique, it became clear that autonomous techniques based on visual observation and dead reckoning were not enough for an accurate determination of position. This was partially solved by the introduction of ground navaids, although the dead reckoning technique still had to be used for those areas that lacked navaid coverage.

2 Natural references - The North

Needless is to say that in order to achieve a proper position awareness, we have to define a clear reference system to which positions are related. From very ancient times, a natural reference system has been used in our planet, taking into account the shape and characteristics of the movement of the Earth. The True North (also referred to as Geographic North) is defined as the intersection point between the surface and the axis of rotation of the Earth. The equator is defined perpendicular to this axis, and it is at the True North that all meridians intersect. We will consider that the True North is a very stable point on the surface of the Earth and that does not change with time.

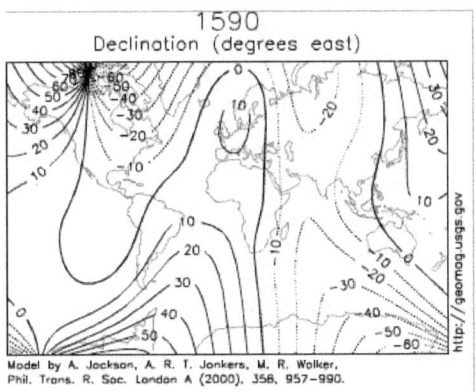

Figure 1.3: Magnetic field declination values for year 1590 © USGS / Public Domain

On the other hand, the Earth is a planet with an internal structure that makes it behave like a huge magnet. This huge magnet has a south pole (towards which all north poles will point) fairly next to the True North. This point on the surface of the Earth towards which a needle of a compass points is defined as Magnetic North, and changes with time.

Thus, we can define, for a given location ($\vec{r}$) and time (t), the angle on a plane formed between the bearing of a needle in a magnetic compass and the True North. This is named *magnetic declination* or *magnetic variation* (δ), and as we have seen, is a function of position and time, $\delta = \delta(\vec{r}, t)$. We will consider δ to be positive when the Magnetic North falls easter than the True North.

Most instruments onboard rely on electromagnetic fields that normally use the Magnetic North as a reference, however, we can´t forget that when we are navigating over the surface of the Earth, we want to move from one geographic point to another one. Corrections imposed by magnetic variation are thus a must and have to be taken into account. Any aeronautical chart will contain information regarding the declination value for a given position and year, in order to let pilots make the necessary corrections. There are areas of the Earth (specially in the vicinity of the Magnetic North) where declination values are very large, imposing restrictions to the use of magnetic-based instruments.

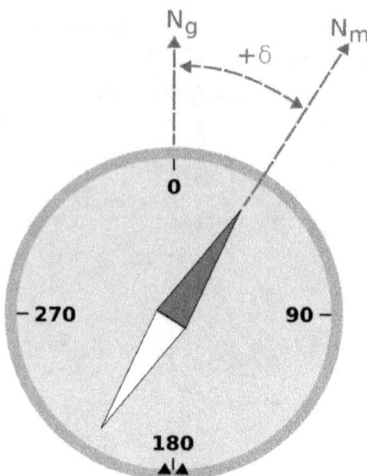

Figure 1.4: Compass showing positive magnetic variation as Magnetic North falls easter than Geographic North © odder / Wikimedia Commons / CC BY-SA 3.0

3   The wind triangle

Air navigation needs to tackle a problem associated to the nature of the movement itself. We have a vehicle (the aircraft) that is flying in the air, but it is through this movement within the air that we want to achieve a certain mission related to ground positions. It is thus important to know not only how the aircraft moves with respect to the mass of air surrounding it (that drives aerodynamic and structural forces in the end), but also how the aircraft moves with respect to the ground (with respect to a different reference system). From very basic kinematics we understand the basic relative velocity equation $\overrightarrow{v_{21}^P} = \overrightarrow{v_{20}^P} + \overrightarrow{v_{01}^P}$. If we name P as a point of the aircraft, being the aircraft the solid $\{2\}$, the Earth the solid $\{1\}$ and $\{0\}$ a reference system fixed to the air. We get that:

- $\overrightarrow{v_{21}^P}$ is the speed of the aircraft with respect to the ground, we will name this Ground Speed $(\overrightarrow{GS})$.

- $\overrightarrow{v_{20}^P}$ is the speed of the aircraft with respect to the mass of air around it, we will name this True Air Speed $(\overrightarrow{TAS})$. In aerodynamics it is normally named $\overrightarrow{U_\infty}$.

- $\overrightarrow{v_{01}^P}$ is the speed point P would have if it belonged to reference system $\{0\}$ with respect to the Earth. It is thus the speed of the mass of air with respect to the Earth, named wind speed $(\overrightarrow{w})$.

We thus have a kinematic definition of the wind triangle, $\overrightarrow{GS} = \overrightarrow{TAS} + \overrightarrow{w}$, that lets us know how the aircraft moves with respect to the Earth, first knowing how it moves with respect to the mass of air and how the air moves with respect to the Earth (wind).

Because of the wind, we have to be aware that not only the magnitudes of vectors $\overrightarrow{GS}$ and $\overrightarrow{TAS}$ are going to change, but also their directions. Aircraft will drift because of lateral wind components. We define **heading or rhumb** as the direction towards which the aircraft longitudinal axis is pointing (be aware that this axis of the aircraft is normally aligned with $\overrightarrow{TAS}$, so we typically say heading comes determined by the direction of the $\overrightarrow{TAS}$ vector). On the other hand we define **track** as the direction towards which the aircraft is flying with respect to the Earth (direction of the $\overrightarrow{GS}$ vector). The angle between the GS and TAS vectors is defined as the **drift angle**.

Taking into account what we just saw in section 2, we can measure angles with respect to two different reference systems (magnetic or geographic). We can thus talk about magnetic heading, true heading, magnetic track or true track.

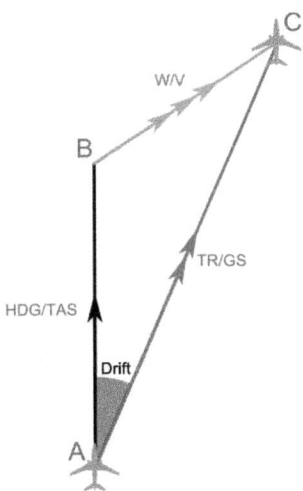

Figure 1.5: Wind triangle components © Abuk Sabuk / Wikimedia / CC BY-SA 3.0

Chapter 2. Meteorology

1 Basic parameters and phenomena

Meteorology is a particular science that covers the behaviour of the mean through which aircraft move (the atmosphere). Understanding and knowing accurately what are the magnitudes of the air surrounding an aircraft is crucial for the safety, efficiency and comfort of the flight. We will briefly talk about different parameters, analyzing their influence on the operations:

- **Wind** - movement of the mass of air with respect to the ground. It comes determined by the direction from which the wind is blowing (in degrees, typically with respect to the Geographic North) and the intensity (expressed in knots).

    - In the *vicinity of an aerodrome*, the direction and intensity of the wind determines the runway in use (to be selected by the tower air traffic controller). Aircraft operations are safer and more stable when facing against the wind, that is, landing and take off maneuvers should take place against the wind. This runway choice criteria leads to lower Ground Speeds while maintaining the required TAS (necessary to produce minimum values of lift), reducing runway occupancy times and increasing the stability of the operations (avoiding high energy situations on the runway). Due to noise abatement procedures, some airports include in their AIP[1] runway configurations to be used with up to 10 knots of tailwind component, which can lead to unstable approaches[2]. Presence of tail or crosswind components increase the probability of a runway excursion; for standard commercial turbofans (e.g. A320 or B737) values over 10 or 35 knots (tailwind or crosswind components, respectively) are typically considered to be critical. A meteorological

---

[1] Aeronautical Information Publication
[2] More information about unstable approaches can be found in http://skybrary.aero

phenomena named **Windshear** is of extreme interest, it consists of sudden changes in the direction or intensity of the wind, leading to sudden changes in the TAS of the aircraft (remember $\overrightarrow{GS} = \overrightarrow{TAS} + \overrightarrow{w}$, and $\overrightarrow{GS}$ does not change instantly since it is related to the kinetic energy the aircraft has). There can be positive or negative windshears (the first leading to an increase of TAS, and thus, an increase of lift, the second one leading to a decrease of TAS and thus of lift). Negative windshear is considered to be of extreme danger when at low altitudes in the vicinity of an aerodrome (typically on short final approach or just after departure), since TAS for these flight phases is close to the stalling speed of the aircraft; thus, sudden changes of the TAS may lead the aircraft to stall[3].

- When *enroute*. Since TAS during the enroute phase is not so close to the stall limits (except for very high altitudes, where the available speed window for the TAS lies within the so called *coffin corner*), changes in the direction and intensity of the wind are not so critical for the safety of the operation (besides a stall at enroute altitudes is easier to recover because of the existence of a larger altitude buffer that can provide time and room for a typical pitch-down stall recovery maneuver). Instead, face or tailwind components have a strong influence on flight times (for a given Mach number for cruise, there is a given TAS, that depending on the wind, delivers a different GS through $\overrightarrow{GS} = \overrightarrow{TAS} + \overrightarrow{w}$). Aircraft operators will thus tend to look for flight routes that maximize tailwind components, leading to shorter flight times and fuel consumptions (as is the case of the North Atlantic jetstream-conditioned route structure).

- **Visibility** - refers to the distance (in meters) up to which it is possible to distinguish a given object (we normally refer to horizontal visibility when generically talking about visibility). Visibility degrades with the presence of fog, clouds, dust, ashes, rain, etc. For a given aerodrome, visibility can be different in each direction. The most critical direction is the one of the runway. Runway visual range (RVR) is defined for each runway direction and is more precisely measured (typically at three different points over each runway). Given the current level of automation and flight instrumentation for commercial aviation, low visibility still conditions flight operations within aerodromes. During the approach phase, horizontal visibility determines whether or not an aircraft can continue the approach after a certain decision point[4]. Lack of visibility raises important hazards for the safety of the operations[5]. Airports that are placed in areas where visibility can be poor do

---

[3]The reader may find some visual examples of windshear following https://www.youtube.com/watch?v=l3SDtn3w_rc

[4]DA for precision approaches, MDA for non-precision approaches

[5]https://www.youtube.com/watch?v=5BvgSS6kBdU shows an incident in which poor visibil-

normally require higher precision navaids for the approach (category of the ILS) in order to lower down the minimums. Since the taxi phase is made manually, relying on visual observation, when visibility is strongly degraded special procedures are implemented limiting the number of operations on the ground, as we will see in section 5. Sometimes, we may also talk about vertical visibility (VV, measured in hundreds of feet).

- **Temperature** - it is used together with pressure to determine air density, which drives aerodynamic forces. Knowing thus the temperature at an aerodrome is essential for all runway performance computations (take off and landing speeds, aircraft configuration and required runway length). Inlet temperature also drives engine efficiency, making thus some vertical levels more efficient than others. Knowing how each day temperature is distributed at each altitude is essential when choosing the required flight level for cruise (we will get deeper into this in Chapter 3). Onboard instruments typically display OAT (Outside Air Temperature, or Static Air Temperature) and TAT (True Air Temperature or Total Air Temperature, T_t). The value of temperature, together with the degree of humidity in the air, gives an estimation of the chances to have fog or ice formation around the aircraft cell. Air humidity is normally depicted using the Dew Point (measured in Celsius degrees), which reflects the temperature at which the vapour in the air would start to condensate (the closer Temperature and Dew Point, the more chances to find fog).

- **Ice** - its presence strongly threatens the safety of an aircraft. Ice formation can change the shape of aerodynamic foils (reducing lift and increasing drag), as well as block flight controls. Pilots tend to avoid ice formation not flying into humid and cold areas (where ice formation is more likely to happen), or through the use of anti-ice systems (i.e. inflating leading-edge boots or spreading bleed hot air around aircraft surfaces).

- **Turbulence** - vertical movements of the mass of air in which the aircraft is flying provoke constant changes in the angle of attack, lifting the aircraft up and down and being normally non-desirable for the comfort of the flight. Severe turbulence encounters may lead to structural or personal damage and if possible, should be avoided. Different meteorological environments favour the appearance of turbulence, being the most common the presence of Cumulonimbus (storm clouds generated through convection processes). For aircraft flying enroute, the presence of turbulence normally leads to aircraft requesting level changes, saturating the available vertical levels and emptying the ones where the turbulence is allocated.

- **Clouds** - Saturated air leads to water condensation, which normally happens at temperatures that are lower than the ones found on the surface of the

---

ity was a strong contributing factor

Earth (giving birth to clouds). There are very many cloud types, depending on their formation processes and altitudes. It is out of the scope of this chapter to give a complete insight into cloud classification, but the most important from the aviation perspective have to be identified and understood. When talking about clouds, we will basically care about three things: how high they are (ceiling of the cloud, or cloud base), how much of the sky they cover (we will measure this in eighths of the sky) and the cloud type (when it can have an impact on the operation). When we talk about the existence of "cloud ceiling", BKN or OVC clouds must exist, otherwise we say "the aerodrome has no ceiling", this is particularly important for the definition of VMC conditions as we will see in section 4.

- Amount of the sky covered - comes determined by the following codes
  * FEW (1-2/8)
  * SCT - scattered (3-4/8)
  * BKN - broken (5-6-7/8)
  * OVC - overcast (8/8)
- Type of clouds - the most dangerous for the operation are the CB (Cumulonimbus) and TCU (Tower Cumulus) because of the presence of ice, hail and strong turbulence inside, that can provoke structural damage and extreme load factors to an aircraft. These types of clouds are formed because of convection processes of hot and humid air that raises up in a very unstable atmosphere (leading later to thunderstorms and associated electrical activity).

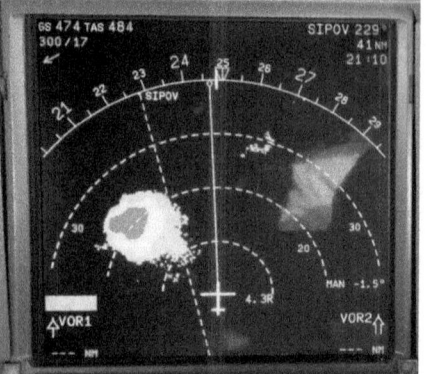

Figure 2.1: Image of an avoided CB and its representation on the ND of the deviated Airbus 330 © Alberto Piquero

2 METAR & TAFOR

Given the importance of meteorological phenomena and its impact on flight operations, it is necessary to have a standard proper mean of communication that assures that all relevant agents are aware of what they are going to encounter. For this purpose, universal coding of meteorology gives birth to METAR, which is a meteorological report of current weather situation at a given aerodrome. TAFOR, on the other hand, stands for forecasts of weather evolution in the vicinity of an aerodrome.

There are also charts and reports indicating enroute relevant phenomena (SIGMET, AIRMET, GAMET) such as the presence of ice, turbulence, wind or temperature for different levels, that fall beyond the scope of this chapter.

ICAO Annex 3 regulates the standard way to code METAR and TAFOR information. Although the complexity of METAR and TAFOR decoding can be very high, we will here stick to the very basics (which are more often used) and that should be easily understood by any agent involved in the flight operation. For the sake of practicality, we will set some examples and decode them. Be aware that wind direction is expressed with respect to the Geographic North, horizontal visibility is expressed in meters and cloud ceiling is expressed in hundreds of feet AGL (above ground level).

- METAR LEBB 031000Z 10005KT 060V150 9999 SCT022 BKN034 20/15 Q1024 NOSIG

  - Bilbao Airport, METAR published on day 03 at 1000Z time, wind blowing from 100 degrees (geographic) with 05 knots of intensity. Wind direction is variable between 060 and 150 degrees (geographic), visibility more than 10 km, scattered clouds at 2200 ft AGL (above ground level), broken clouds at 3400 ft AGL, temperature 20ºC, dew point 15ºC, QNH 1024 mb. No significant changes expected.

- METAR EGLL 131220Z 26019KT 9999 FEW030 SCT042 13/03 Q1012 TEMPO SHRA

  - London Heathrow Airport, METAR published on day 13 at 1220Z time, wind blowing from 260 degrees with an intensity of 19 knots, visibility more than 10 km, few clouds at 3000 ft AGL, scattered clouds at 4200 ft AGL, temperature 13ºC, dew point 03ºC, QNH 1012 mb, temporarily expect shower rain.

- METAR GCLP 131230Z 01025G35KT 340V040 9999 FEW020 24/16 Q1018 NOSIG

  - Las Palmas Airport, METAR published on day 13 at 1230Z, wind blowing from 010 degrees with intensity of 25 knots gusting up to 35 knots, direction variable between 340 degrees and 040 degrees, visibility more than 10 km, few clouds at 2000 ft AGL, temperature 24°C, dew point 16°C, QNH 1018 mb, no significant changes are expected.

- TAFOR EGLL 131057Z 1312/1418 26015KT 9999 SCT040 TEMPO 1312 / 1320 25018G28KT 7000 SHRA BECMG 1410/1413 17012KT 8000 -RA BKN014 PROB40 TEMPO 1412/1418 4000 RA BKN009

  - London Heathrow Airport, TAFOR published on day 13 at 1057Z, forecast valid between day 13 at 1200Z and day 14 at 1800Z, wind blowing from 260 degrees with an intensity of 15 knots, visibility more than 10 km, scattered clouds at 4000 ft AGL, temporarily, between day 13 at 1200Z and day 13 at 2000Z, wind blowing from 250 degrees with intensity of 18 knots gusting up to 28 knots, visibility of 7000 meters, shower rain, becoming from day 14 at 1000Z to day 14 at 1300Z wind from 170 degrees with intensity of 12 knots, visibility 8000 meters, light rain, broken at 1400 ft AGL, with probability of 40%, temporarily between day 14 at 1200Z and day 14 at 1800Z, visibility 4000 meters, rain, broken clouds at 900ft AGL.

3 VFR & IFR. Z and Y Flight Plans

VFR stands for Visual Flight Rules and refers to the navigating procedures that the pilot can use to conduct the flight. VFR flights navigate basing their position awareness on visual observation (rivers, towns, mountains, etc). This does not mean that VFR flights do not have instruments, of course they have (anemometers, variometers, etc...), what it means is that their navigation performance is based on visual observation. On the other hand, IFR stands for Instrument Flight Rules and refers to navigations that take place based on lateral awareness instruments (such as VORs, NDBs, ILS, or even RNAV navigation). IFR flights can only take place if the aircraft is properly equipped (has the necessary onboard equipment to receive and display signals from different navaids), there exists ground infrastructure that allows the navigation (navaids), and the crew is trained and certified in the use of these instruments. Take into account that an IFR equipped aircraft can be flying VFR, but not the other way around. IFR had its origin when the incipient aviation industry realized that depending on the visual conditions, navigating using only outside visual reference could be unsafe.

Before performing a certain flight, the crew has to fill in the so named Flight Plan, a document in which the crew writes down the intentions for the flight. In this document, the pilot must specify whether the flight will be done according to visual or instrument flight rules.

Not all airports are prepared to support IFR flights (they may not have navaids or published instrument procedures to guide aircraft during the approach). Y or

Z flight plans (instead of V or I, for VFR or IFR flight plans) let pilots change the rules according to which they are flying in the middle of the flight:

- Z flight plans start as VFR and later change to IFR (e.g. departing from a small visual field, and landing in a major instrument airport).

- Y flight plans start as IFR and later change to VFR (e.g. departing from a navaid-equipped airport and landing in a visual field)

4 VMC & IMC

VMC stands for Visual Meteorological Conditions and refers to the minimum weather conditions under which visual flights can take place (VMC required to fly VFR). Depending on the country, altitude, airspace class or even speed of the vehicle (helicopters), VMC conditions are defined differently. In Europe this comes determined by SERA (Single European Rules of the Air) articles 5001 and 5005, defining different requirements for each scenario, providing minimum values of visibility, cloud ceiling and minimum distance that has to be preserved from the clouds. Since METARs are published reporting weather in the vicinity of aerodromes, it is of particular importance the definition of VMC applied for the vicinity of an aerodrome (ATZ, Aerodrome Transit Zone), that requires at least:

- Minimum cloud ceiling (BKN or OVC) equal or above 1500ft AGL

- Visibility equal or above 5000 meters

IMC stands for Instrument Meteorological Conditions. If VMC conditions do not apply, then we say IMC conditions apply. VFR traffic is not allowed under IMC conditions (only IFR traffic is permitted). Be aware that IFR traffic can fly no matter whether the situation is VMC or IMC, while VFR can only fly under VMC conditions.

5 LVP

LVP[6] stands for Low Visibility Procedures, and are typically applied when IMC conditions degrade extremely, reaching the so-called LVC (Low Visibility Conditions). LVC threshold values and LVP are defined particularly for each airport (in their AIP) and normally consist of a series of measures that are taken to avoid ground collisions between aircraft and between aircraft and vehicles. Some of the measures that they may include are:

- Activation of Stop Bars to prevent runway incursions

---

[6]A video of a departure under LVC can be found in https://www.youtube.com/watch?v=-Sa_MUzrYt4

- Restriction of ground vehicles, which normally remain within the apron (avoiding taxiways and runways)

- Air Traffic Control specific procedures (e.g. only one aircraft on the maneuvering area at a time), which are more restrictive if the airport lacks ground surveillance radar equipment

Chapter 3. Aircraft Instruments

Not long after first aircraft started flying, came the need to assist pilots to better know about the attitude of their aircraft. Instruments that could help pilots navigate through clouds, fog, rain or just during the night became more a more necessary for an aviation seeking for growth. New aircraft could fly faster and make more demanding maneuvers that pilots' senses could not easily decipher, getting disoriented. One of the pioneers developing first instruments was Jimmy Doolittle, he understood that independent instruments that could help pilots make decisions were key for aviation success and contributed to the development of first artificial horizons. Since then, many pilots have been instructed to trust their instruments instead of their senses. Having good, understandable and reliable instruments is key for the safety of an operation and is still today a matter of concern when we see accidents in which failure of sensors or wrong interpretation of instrumentation has led to damage or loss of lives.

1 Units

Despite of the important value non-dimensional analysis shows to any engineer candidate, we can´t forget that here we are talking about real flying and true sense of reality. It is thus a need to have true units with which pilots are familiarized. The following units are extensively used in the aviation industry, mainly because of the important success American and British industries had during the evolution of aviation.

- Horizontal Distance - Nautical Miles (nm[1]) being 1 NM = 1852 meters

- Vertical Distance - Feet (ft) being 1 FT = 30.48 cm

- Horizontal Speed - Knots (kts) being 1 Knot = 1 NM per hour

- Vertical Speed - Feet per minute (FPM or ft/min)

---

[1]A nautical mile is also defined as the length of one minute of arc of an earth´s great circle

2 Altimetry

2.1 Introduction

When talking about altimetry we talk about determining the vertical position of the aircraft (with respect to a given point or surface). Out of the many ways to determine aircraft vertical position (through visual observation, GPS-determined, etc), we will focus on those which are generally used in aviation: radioaltimetry and barometry.

Determinining the vertical position of an aircraft is vital for a safe and optimum operation, and thus the whole altimeter system (starting from sensors to the final read output) is crucial. Here we list some of the reasons for which having a right vertical position awareness is essential:

- Avoidance of Controlled Flight Into Terrain events (CFIT) - Fatal accidents and incidents occur when a lack of vertical position assesment drives the pilots towards a situation of controlled flight that ends up hitting the ground or obstacles, especially in the presence of fog, clouds or during the night.

- Vertical separation with other aircraft - as we will cover later, vertical separation is provided to let different aircraft be at the same lateral position at the same time.

- Weather related events - forecast of ice or turbulence at given flight levels, requiring the crew to avoid them asking for a different vertical flight level.

- Efficiency - Air magnitudes such as temperature, pressure and density determine engine and aircraft performance. These magnitudes evolve as a function of altitude, and thus looking for certain air magnitudes is often done reaching the altitude at which they are expected.

2.2 Altitude, height and elevation

As previously stated, our aim is to determine our vertical position with respect to a given point or surface. The following three concepts are used in Air Navigation.

- **Altitude** - when referring our vertical position with respect to mean sea level (MSL). We know we are reading altitudes on a navigation chart when reading AMSL (Above Mean Sea Level) - *(i.e. fly over PND at 5000ft AMSL or above)* - or when no other indication is provided.

- **Height** - when referring our vertical position directly with respect to the ground. We know we are reading heights on a navigation chart when reading AGL (Above Ground Level) - *(i.e. maintain always 1000ft AGL, meaning we have to maintain 1000ft height over the ground)*

- **Elevation** - when referring the vertical position of an aerodrome or an obstacle (fixed point on earth) to mean sea level *(i.e. LEMD elevation is 2010ft)*

Figure [3.1] summarizes these three concepts.

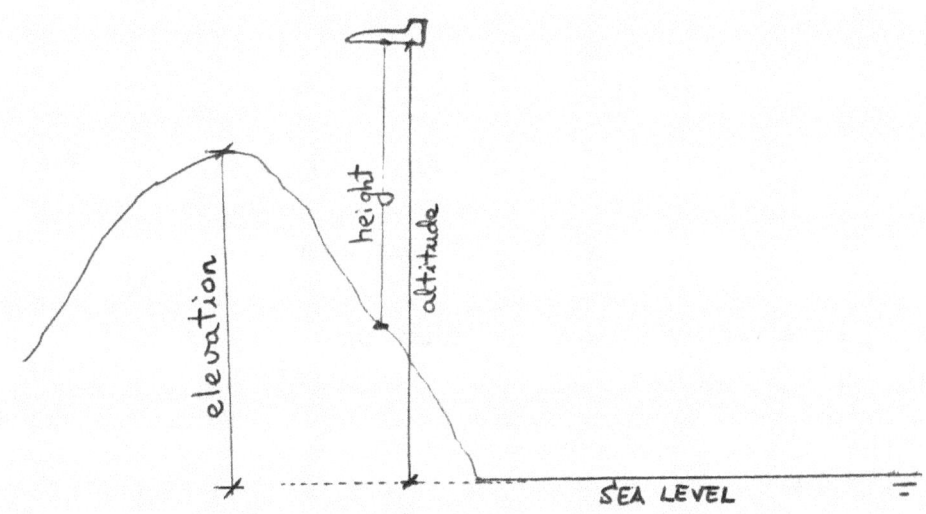

Figure 3.1: Definitions of altitude, height and elevation

2.3 Radioaltimeter

Radioaltimeter is an instrument displaying height. It works receiving input from a system of transmitters and receivers that send radio signals vertically downwards, expecting an echo. By measuring the time it takes for the radio signal to go down, bounce against an obstacle, and come back, the radioaltimeter determines the distance between the aircraft and the obstacle. This process is summarized in Figure [3.2].

Radioaltimeters do normally not range further than 2500ft, being thus useful only when the aircraft is close to the ground (take off and landing). They play a key role in the functioning of Ground Proximity Warning Systems[2] and pilot position awareness during the approach, especially regarding Decision Height (DH)[3] awareness.

---

[2]An interesting malfunctioning case of radioaltimeter that partially led to an accident can be checked in http://avherald.com/h?article=41595ec3/0070

[3]Decision Height is the lowest height an aircraft can get during a precision approach without having visual contact with the runway

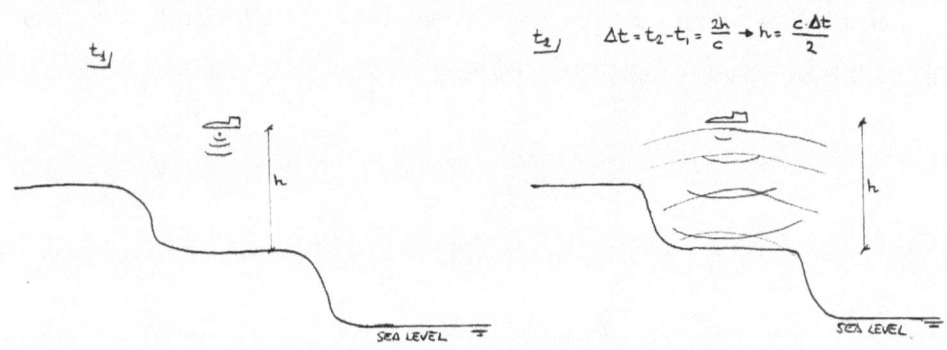

Figure 3.2: Radioaltimeter

2.4 Pressure evolution with altitude

The following reasoning is valid within the troposphere, that is, from sea level up to the tropopause, which for ISA is established at 11000 meters of altitude. From a simple one-dimensional equilibrium equation of forces for the volume of air depicted on Figure [3.3], assuming $p = p(z)$ we can state that

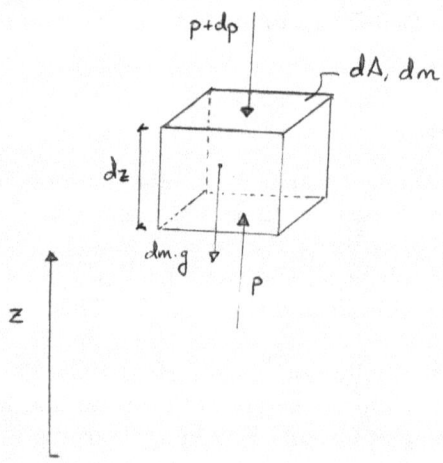

Figure 3.3: Infinitesimal volume of air

$$pdA - (p + dp)dA - dmg = 0$$

and since $dm = \rho dV$ and $dV = dzdA$, thus

$$p \cdot dA - (p + dp) \cdot dA - \rho \cdot g \cdot dA \cdot dz = 0$$

so

$$dp = -\rho g dz$$

being $\rho = \rho(z)$ since $\rho = \frac{p}{RT}$ and we assume a linear decay of temperature with altitude until the tropopause $T = T_{SL} - \alpha z$ being $\alpha = 6.5\,10^{-3}\frac{K}{m}$ and $T_{SL} = 288\,K$ (sea level temperature for a standard ISA day)
Thus

$$\frac{dp}{p} = \frac{-g dz}{R(T_{SL} - \alpha z)}$$

And integering from conditions $p = p_{SL}$ when $z = 0$, we finally get

$$Ln\frac{p}{p_{SL}} = \frac{g}{\alpha R} Ln\frac{T_{SL} - \alpha z}{T_{SL}} \rightarrow p = p_{SL}\left(1 - \frac{\alpha z}{T_{SL}}\right)^{\frac{g}{\alpha R}} \tag{1}$$

For the International Standard Atmosphere (ISA) values are $p_{SL} = 1013mb$ and $T_{SL} = 288K$. However, as we can see on Figures [3.4] and [3.5], different values of T_{SL} (i.e. ISA deviations) and p_{SL} (different QNH values, as we will see in the next section) lead to different pressure vs altitude distributions according to equation (1).

2.5 Determination of altitude

As we just saw in Section 2.4, for a standard day, considering sea level values of p_{SL} and T_{SL}, we know how pressure evolves with altitude, and since Equation (1) is a monotonically decreasing function, we can also define $z = z(p)$, which is the main purpose of barometric altimetry. Thus, for a day with given boundary conditions (at sea level) it will work:

$$z = \frac{T_{SL}}{\alpha}\left[1 - \left(\frac{p}{p_{SL}}\right)^{\frac{\alpha R}{g}}\right] \tag{2}$$

However, ISA boundary conditions do often not apply (temperature and pressure at sea level different from standard T_{SL} and p_{SL} values). Plus, as we have previously seen, we are not always interested in reading on our altimeter our vertical distance with respect to mean sea level. We will then define a changing reference surface with its associated pressure value (z_{ref}, p_{ref}).

23

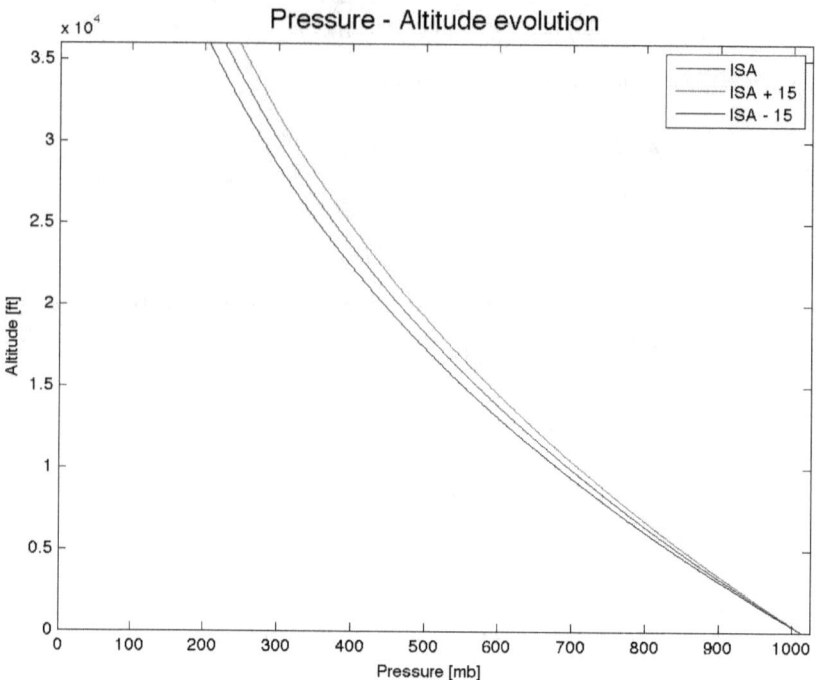

Figure 3.4: Pressure evolution with altitude for different T_{SL} being $p_{SL} = 1013mb$

We will then measure our vertical position with respect to this z_{ref} surface, being

$$\Delta z = \Delta z(p) = z - z_{ref} = \frac{T_{SL}}{\alpha} \left[\left(\frac{p_{ref}}{p_{SL}} \right)^{\frac{\alpha R}{g}} - \left(\frac{p}{p_{SL}} \right)^{\frac{\alpha R}{g}} \right] \tag{3}$$

where the only needed input is the static outside pressure p that is obtained through the static ports of the aircraft and where p_{ref} is selected by the pilot on the altimeter′s Kollsman Window (see Figure [3.6]) depending on what he wants to read, as we will now study. Altimeters are calibrated using ISA values for T_{SL} and p_{SL}, which leads to errors when real conditions do not match ISA. That is, when talking about altimeter readings one must take into account that in equation (3) $T_{SL} = T_{SL,ISA}$, $p_{SL} = p_{SL,ISA}$ and $\alpha = \alpha_{ISA}$.

The following reference isobars are typically used for altimeter setting:

- **QNH** - is the pressure value that there would be at sea level for a particular position and time if we could drill a hole into the ground and reach sea level. That is, it is the value p_{ref} for $z_{ref} = 0$. QNH is normally computed

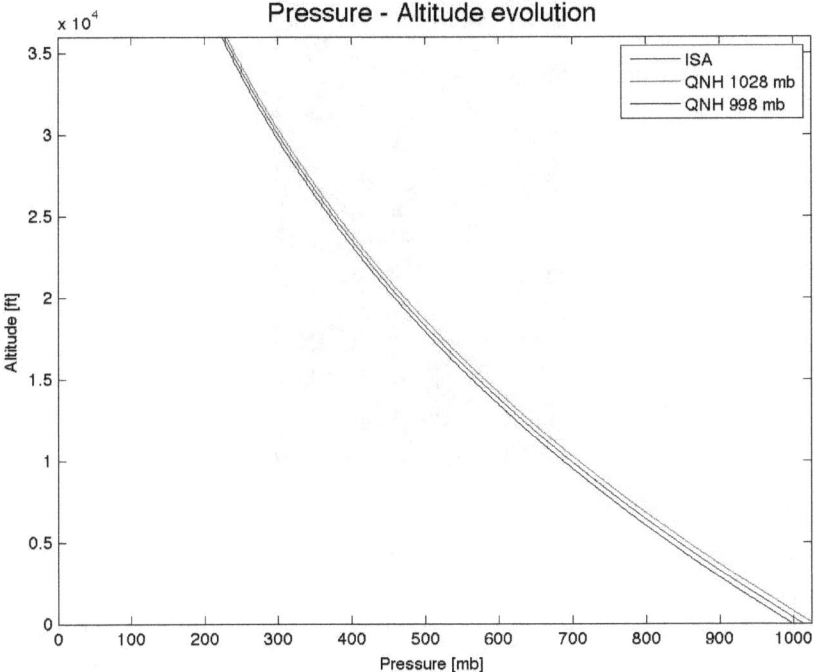

Figure 3.5: Pressure evolution with altitude for different p_{SL} being $T_{SL} = 288K$

knowing p for a particular z and solving for p_{ref} in equation (3). When on the ground, if an altimeter is adjusted so that we can read the elevation at which we are, the Kollsman window will indicate the QNH for that position and time. When selecting QNH, altimeter reads altitudes.

- **QFE** - normally referred for a given runway threshold, it is the pressure at that particular point. That is, it is the value p_{ref} for $z_{ref} = z_{threshold}$. QFE is directly retrieved from any pressure sensor on the ground. When over the given threshold, if altimeter has QFE in the Kollsman window, the altimeter is reading heights. Be aware that one runway will have two associated QFE values, one for each runway threshold.

- **QNE** - $p_{ref} = 1013\,mb$ or $29.92\,inHg$. Altimeter reads Flight Levels (FLs). Flight Level reflects the vertical distance in hundreds of feet to the $1013\,mb$ isobar (e.g. FL226 means 22600ft above the $1013\,mb$ isobar).

Standard cartography tends to show elevation of obstacles and terrain with respect to mean sea level, that is why in the vicinity of airports and when flying close to the ground, it is interesting to read altitudes (local QNH selected). However, when

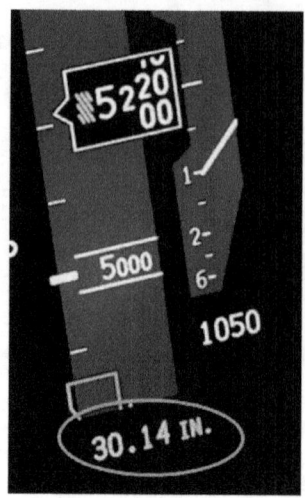

Figure 3.6: Kollsman window for a barometric altimeter using inHg units
© 737Channel

reaching higher levels, aircraft may encounter other aircraft that have departed from different airports, all of which initially selected their local origin QNH on their altimeters. If spacing between different aircraft wants to be established, instructions to maintain certain vertical positions are given, being thus important that all aircraft have the same z_{ref} (all measuring their vertical distances with respect to the same surface), which gives birth to QNE. QNE is a standard reference isobar that all pilots must select as p_{ref} when climbing through the Transition Altitude.

2.6 Transition Altitude, Level and Layer

- **Transition Altitude (TA)**- it is the altitude, published for each departure airport, at which a change of p_{ref} has to be made, from QNH to QNE when climbing. Below TA we fly altitudes, above TA we fly Flight Levels. TAs are normally published on the SID charts of a given airport.

- **Transition Level (TL)** - it is a Flight Level (FL) placed at least 1000ft above the TA at which a change of p_{ref} has to be made, from QNE to QNH when descending. Below TL we fly altitudes, above TL we fly Flight Levels. Depending on the QNH value, TL varies to ensure minimum 1000ft separation between TA and TL. Available TL happen every 500ft (*i.e. for a TA of 6000ft, typical TL range from FL065 to FL085, depending on QNH value*).

- **Transition Layer** - it is the layer between the Transition Altitude and

Transition Level. Minimum width for Transition Layer is 1000ft. No aircraft is allowed to level off within the Transition Layer, it is designed as a vertical region where aircraft do only transition (climbing or descending, but never levelling off).

Figure [3.7] shows these three concepts.

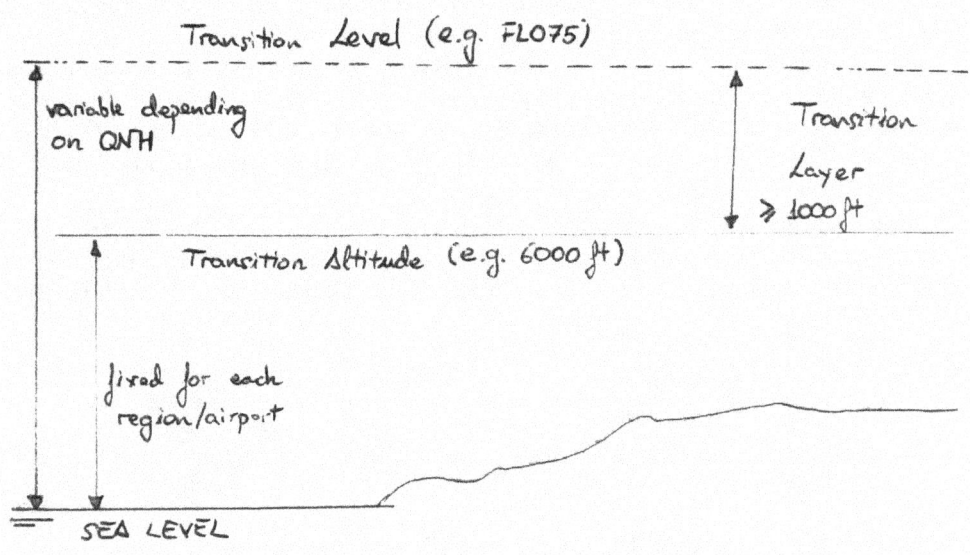

Figure 3.7: Transition Altitude, Transition Level and Transition Layer

2.7 Vertical separations - RVSM

One way to let different aircraft fly simultaneously the same lateral route is by assigning each one a different Flight Level. This is called vertical separation and is commonly used. On the same day, depending on vertical temperature distribution and aloft winds, most aircraft tend to request very similar FLs, thus, they are normally allowed to fly close to each other (laterally) around their optimum requested FL.

The question to be addressed is then how close to each other (vertically) may we let aircraft fly? this will obviously depend on the accuracy of their altimeters and their ability to maintain an instructed FL. Accuracy of altimeters decays with altitude, while the ability to maintain a given altitude or FL is normally dependant on the availability of an operating autopilot. These limitations traditionally imposed wider vertical separations at higher levels.

Originally, we talked about CVSM (Conventional Vertical Separation Minima), an structure of airspace that would separate aircraft every 1000 ft until FL290, over which, the required minimum vertical separation would increase to every 2000ft (over FL290, aircraft could only fly at FL310, FL330, etc...). Current state-of-the-art aircraft tend to find their optimum FLs between FL340 and FL400, just at the core of these 2000ft separation restrictions. Increasing demand during the 90s exceeded existent capacity (strongly related to the number of available FLs), thus, it came evident the need to increase the number of available Flight Levels, reducing aircraft vertical separation when possible.

A Reduced Vertical Separation Minima (RVSM[4]) concept was developed under which minimum 1000ft separation requirement would extend (for certified aircraft and crew) up to FL410. This multiplied the number of available FLs (FL300, FL320, etc), fostering capacity. However, current accuracy of altimeters still require 2000ft vertical separation above FL410.

Aircraft or crew which are not RVSM certified can not fly higher than FL290, unless exempted (i.e. military aircraft under some circumstances), in which case a 2000ft vertical separation has to be provided. Table [3.1] summarizes both CVSM and RVSM concepts.

Available FLs	CVSM	RVSM
Up to FL290	every 1000ft	every 1000ft
From FL290 up to FL410	every 2000ft	every 1000ft
From FL410 up to UNL	every 2000ft	every 2000ft

Table 3.1: CVSM and RVSM concepts

2.8 Level Busts

As already explained within section 2.7, vertical separation between different aircraft can only be achieved if procedures and systems enable stable altitude control and monitoring. Any violation regarding the assigned altitude (or FL) of an aircraft is known as Level Bust.

Level Busts currently represent an important amount of the total number of aviation incidents (still remaining many of them unreported) due to several reasons such as communication misunderstandings (between pilots and ATC), human selection error, wrong altimeter setting, etc.

Level Busts may lead to AIRPROX events, triggering last safety-net systems such as TCAS[5].

---

[4]See ICAO Doc 9574
[5]Traffic Colision Avoidance System

2.9 Variometer

Variometers measure the rate at which altitude changes in time (normal units are feet per minute, although gliders and soviet countries have traditionally measured vertical speed in meters per second). This instrument basically tries to compute the first derivative of z, that is $\frac{dz}{dt}$, using static ports that deliver instant static pressure into a capsule that has a leak so that the air surrounding the capsule has the static air pressure that there was a calibrated Δt before. By measuring this difference, $p(t) - p(t - \Delta t)$, the instrument computes $\frac{p(t)-p(t-\Delta t)}{\Delta t}$ and displays $\frac{z(t)-z(t-\Delta t)}{\Delta t}$.

Figure [3.8] shows a basic description of an analogical variometer.

Diaphragm

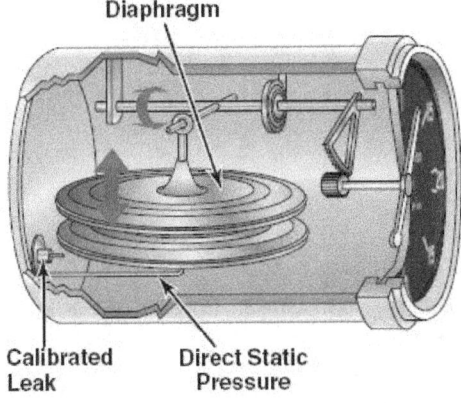

Calibrated Leak   **Direct Static Pressure**

Figure 3.8: Basic variometer mechanism © Wikimedia Commons / Public Domain

3 Anemometry

Having a valid indication of the speed of the airplane is essential to derive and know different aspects of the flight, aerodynamic forces, structural stresses as well as air navigation parameters (such as time estimates to different navigation points), which are all going to depend on the speed of the aircraft. As we will see, measuring the speed of the aircraft with respect to the mass of air around (airspeed) is not as straightforward as it may look. Depending on what and how we measure impact-air magnitudes, we will come up with different speed definitions and values.

Anemometry is strongly related to measuring impact air magnitudes, basically stagnation pressure, static pressure and static temperature. Sensors such as Pitot-Tubes, Static Pressure Ports and Termometers are essential to retrieve these magnitudes.

3.1 Stagnation pressure (p_t). TAS. CAS. IAS. EAS.

Stagnation pressure (also known as total pressure or pitot pressure) is defined as the static pressure we would get if we decelarated a fluid without the use of gravity nor viscous forces, not adding heat from velocity U_∞ to null velocity, which is exactly the proccess air experiments when an aircraft is flying at U_∞ and air hits a pitot tube. From now on, we will name the airspeed of the aircraft (speed with respect to the mass of air) as TAS (True Air Speed), instead of U_∞.

For incompressible flows ($M \ll 1$) we can compute the value of $p_t = p + \frac{1}{2}\rho TAS^2$, so $TAS = \sqrt{\frac{2(p_t-p)}{\rho}} = \sqrt{\frac{2\Delta p}{\rho}}$ where $\Delta p = p_t - p$ is the difference between the stagnation and static air pressure.

For compressible flows (non-negligible Mach number) we have

$$p_t = p\left(1 + \frac{\gamma-1}{2}M^2\right)^{\frac{\gamma}{\gamma-1}} \tag{4}$$

And since we know that $M = \frac{TAS}{\sqrt{\gamma RT}}$ and $p = \rho RT$, doing some algebra we can easily get that

$$TAS^2 = \frac{2\gamma}{\gamma-1} \cdot \frac{p}{\rho} \cdot \left[\left(\frac{\Delta p}{p} + 1\right)^{\frac{\gamma-1}{\gamma}} - 1\right] = f(\Delta p, p, \rho) \tag{5}$$

which basically tells us that in order to be able to compute our TAS, we not only need to measure the increase of pressure because of the impact (Δp) but also the static value of the pressure itself (p) and the air density (ρ) or another equivalent value (like, for example, T, since having p and T we can get ρ).

Since a mechanic anemometer (see Figure [3.9]) only measures Δp (substracting static pressure to stagnation pressure), fixed values for the other variables had to be chosen in order to solve equation (5). This gives birth to CAS (Calibrated Air Speed), which is only a function of Δp, giving standard sea level values to all other thermodynamic magnitudes that play a role in equation (5). Some authors say that TAS is CAS corrected by density effects, which is another valid interpretation of these equations.

$$CAS^2 = \frac{2\gamma}{\gamma-1} \cdot \frac{p_{SL}}{\rho_{SL}} \cdot \left[\left(\frac{\Delta p}{p_{SL}} + 1\right)^{\frac{\gamma-1}{\gamma}} - 1\right] = f(\Delta p) \tag{6}$$

Since instruments have errors, we will name IAS (Indicated Air Speed) to the speed the pilot actually reads on the anemometer. CAS is then IAS corrected by instrument error.

Although it is not relevant for air navigation, many structural engineers talk about EAS (Equivalent Air Speed), which is just the speed an aircraft would need to fly at sea level to have the same value of dynamic pressure, that is:

$$\frac{1}{2}\rho TAS^2 = \frac{1}{2}\rho_{SL} EAS^2 \rightarrow EAS = \sqrt{\frac{\rho}{\rho_{SL}}} TAS \tag{7}$$

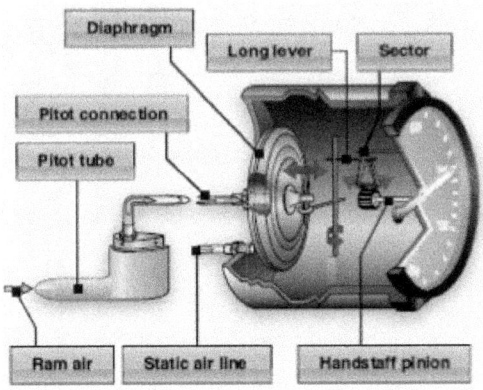

Figure 3.9: Mechanic structure of a simple anemometer © Wikimedia Commons / Public Domain

3.2 Importance of speed indication

Recent tragic events[6] in which unreliable speed indications derived from ice-blocked pitot tubes led to pilots being clueless about what was happening, have opened the debate about the right understanding of speed indications and the sources from which each display takes different magnitudes. We have to understand that the process follows the following scheme.

$$\Delta p \to IAS \to CAS \to TAS$$

$$Inertial\ Navigation\ Systems\ (INS)\ supported\ by\ GNSS \to GS$$

For an ISA day, Figure [3.10] depicts how the relationship between CAS and TAS changes for different altitudes.

That is, the pitot-tube and static port sources deliver a value for Δp that forces the expansion of a mechanical capsule, moving a needle or display that shows the IAS. If corrected taking into account instrument or visualization errors, this would give us the CAS, which correcting for the current true air conditions (T or ρ) will give us the True Airspeed at which the aircraft is moving with respect to the mass of air (responsible for the generation of aerodynamic forces and loads). Using wind information, we could solve the wind triangle and finally get our Ground Speed (GS), which is what matters most for air navigation (flight tracking and conflict resolution).

Normally GS is obtained using INS, and the wind triangle equation is used onboard to solve for the wind, being $\overrightarrow{w} = \overrightarrow{GS} - \overrightarrow{TAS}$. This information is typically

---

[6]AF447, an A330 flight from Rio to Paris in 2009 or QZ8501, an A320 flight from Juanda to Singapore in 2014; both with fatal consequences

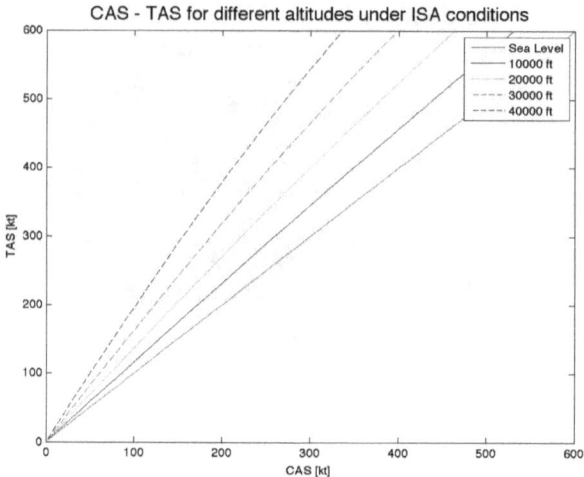

Figure 3.10: TAS vs CAS relationship at different altitudes for ISA conditions

displayed on the upper left corner of the navigation display (see Figure [3.11]) and can only be fully retrieved when the aircraft is airborne and thus drifts, being $\overrightarrow{TAS}$ aligned with the longitudinal axis of the aircraft and $\overrightarrow{GS}$ not forced to follow certain ground trajectory aligned with the longitudinal axis of the plane (as may happen when taxing on the ground or during take off roll).

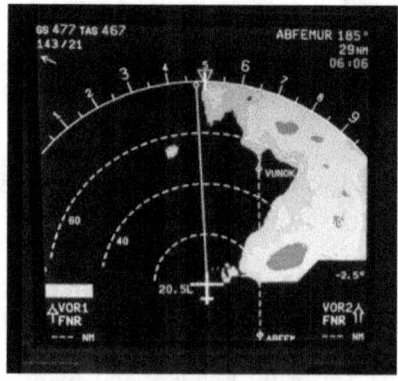

Figure 3.11: Navigation Display showing wind blowing from 143 degrees with an intensity of 21 knots in an Airbus 330 © Alberto Piquero

Chapter 4. The ATM/CNS Concept

Air navigation must not only care about how each individual aircraft determines its position and intended route, but also about the whole system as a safe environment in which a network of flights take place. As a result, juridic, operative, and technical support frameworks are established. The technical and operative frameworks must supply:

- Information that turns out to be relevant, prior to departure, with a content based on the the valid meteorology reports, operative procedures and navaid limitations

- Tactical support to pilots related to possible modifications in the trajectory of the flight, specially to avoid potential conflicts with other aircraft while ensuring separation with terrain

- Radioelectric infrastructure required for the determination of position, that is, navigation aids

These three items have constituted the basic pillars of what is referred to as the *air navigation system*. The *technical and operative* frameworks that conform the system give shape to the so-called ATM/CNS concept. CNS (Communication, Navigation and Surveillance) corresponds to the required technical means which are externally required by the aircraft to fulfill a mission from the air navigation perspective, while ATM (Air Traffic Management) refers to the set of tactical actions, operational procedures, airspace definitions and required aeronautical information. The *juridic* framework falls within a different subject (Air Transport) and studies, among others, the set of international law requirements that have to be commonly implemented for a safe and efficient deployment of the technical and operative means that we will cover along the following lines. It is through this juridic framework that the specifications and standards for each of the required services and infrastructures are established, as well as the associated liabilities.

1 ATM Concept

The Air Navigation Services (ANS) include all the services that are provided to airspace users (e.g. airlines) such that flight operations can be safely (and efficiently) performed. These services are provided at each airspace volume by the assigned country for this responsibility. The provider of these services, which can be either government departments, state-owned firms, or even private organizations, are referred to as Air Navigation Service Providers (ANSPs). Figure [4.1] presents a scheme of the different ANS services.

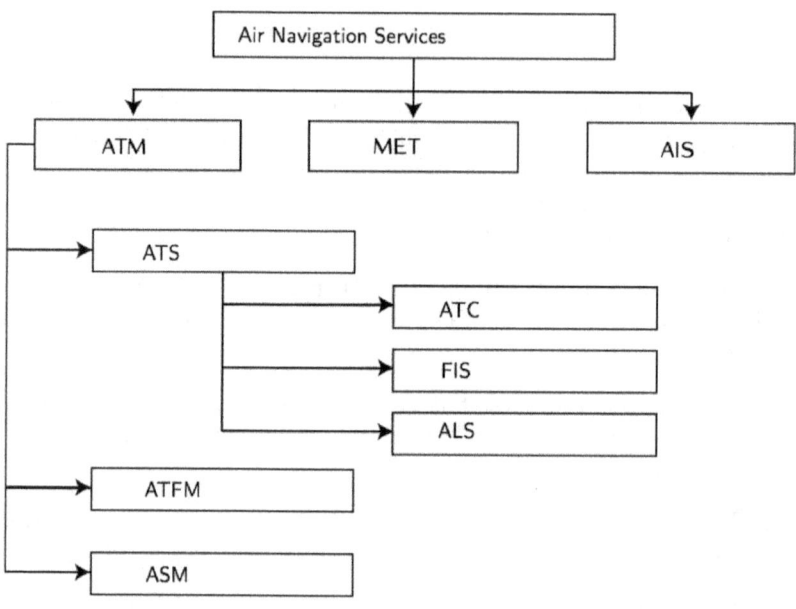

Figure 4.1: The Air Navigation Services

1.1 Aeronautical Information Service (AIS)

The Aeronautical Information Service (AIS) is defined *as the service which provides the necessary information to ensure aeronautical operations develop with safety, regularity, economy and efficiency.* All this information must be made public and properly distributed.

The information included in the AIS is composed of:

- Aeronautical Information Publication (AIP) - The AIP is a basic aeronautical information manual. It contains permanent information and long-term changes. It is used essentially for air navigation and airport operations. It

34

contains information on the definition and availability of routes, procedures, navigation charts, etc, both for airports and en-route areas.

- AIP Amendments (AMDT) - Regular amendments (AMDT) include permanent changes and editorial corrections of the AIP. The most important AMDT is the AIRAC cycle, an AIP update which has to be is issued every 28 days.

- Supplements (SUP) - These complement or vary the information contained in the AIP. They contain temporary information that requires extensive texts and/or explanatory graphics.

- Notice to Airmen (NOTAM) - they are notices distributed by means of telecommunication containing information concerning the establishment, condition or change in any aeronautical facility, service, procedure or hazard. The timely knowledge of their content is essential to personnel concerned with flight operations (i.e. pilots and air traffic controllers). NOTAMs are issued by national authorities for a number of reasons, for example:

    - Hazards such as air-shows, parachute jumps or glider activities (see Figure [4.2])
    - Closed runways, taxiways, etc
    - Military exercises with associated airspace restrictions
    - Limitations in the use of certain navaids

- Aeronautical Information Circulars (AIC) - is a notice containing information that does not qualify for the origination of a NOTAM nor for its inclusion in the AIP, but which relates to flight safety, air navigation, technical, administrative or legislative matters.

Further information regarding the content that must be part of the AIS can be consulted in ICAO's Annex 15.

1.2 Meteorological Service (MET)

The MET Service for international aviation is provided by meteorological authorities designated by states through the use of standardized MET products and services delivered in accordance with ICAO Annex 3. Each State also establishes a suitable number of meteorological offices, which are: Aerodrome Meteorological Offices (AMOs), Meteorological Watch Offices (MWOs) and Aeronautical Meteorological Stations (AMSs). These services were established on the prevailing state-of-the-art available in the 1960s, and consist mainly in messages which are distributed using coded information. Basically:

- METAR/TAF: Aerodrome MET conditions/forecast

```
(D0459/17 NOTAMN
Q)LECM/QRTCA/IV/BO /W /000/150/4220N00527W005
A)LECM B)1703010700 C)1703311850
D)SR-SS
E)TEMPORARY RESTRICTED AREA FOR PJE ACTIVATED
WI 05NM RADIUS OF 421959N 0052707W
LEON/LOS OTEROS
F)SFC G)FL150)
```

Figure 4.2: Example of a NOTAM message containing information about parachuting activities over a given area

- SIGMET: En-route significant weather advisory

- AIRMET/GAMET: En-route weather phenomena (typically related to lower levels, for general aviation users)

Moreover, weather forecasts which include wind, turbulence and convective information for the en-route phase of the flight are prepared by World Area Forecast Centres (WAFCs). This ensures the provision of high-quality and uniform forecasts for flight planning and flight operations.

1.3 ATM Services

ANSPs must have technical capabilities to develop and support, as we will see in section (2) a technical CNS infrastructure, but on the other hand, a highly structured organization with high skilled people is required, giving shape to the operational support to provide transit, communication, and surveillance services. This operational infrastructure is referred to as Air traffic Management (ATM). ATM is about the process, procedures, and resources which come into play to make sure that aircraft are safely guided in the skies and on the ground. If we consider a time-horizon line starting with airspace design and management, and ending with the flight taking place, we can identify three system levels:

- AirSpace Management (ASM) - Strategic level

- Air Traffic Flow and Capacity Management (ATFM) - Pretactical level

- ATS Services (including Air Traffic Control) - Tactical level

All these three services will be covered along the lines of this book with one single chapter devoted to each of them. However, we will here state an introductory short description.

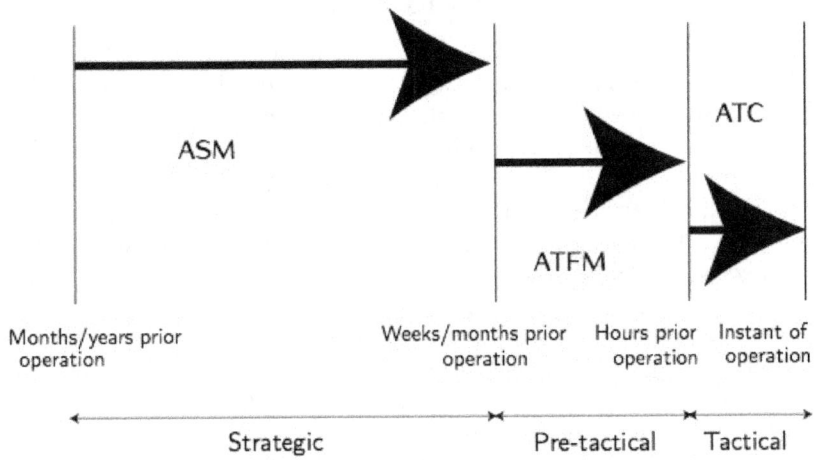

Figure 4.3: Different system levels and associated timescale

1.3.1 Airspace Management (ASM)

The first layer of the ATM system, the so-called Airspace Management (ASM), is performed at a strategic level, months or even years prior to aircraft operation. ASM is an activity which includes airspace modeling and design. As aircraft fly in the sky, they follow specified routes conformed by waypoints, with associated altitude restrictions. When coming out from their origin, they follow departure procedures (commonly known as SIDs, Standard Instrument Departures). When getting closer to their destination, they follow arrival procedures (STARs, Standard Arrivals) and end up flying approach procedures to the threshold of the landing runway (VOR, ILS or even RNAV approach procedures). The planned route to be followed by an aircraft is chosen by the airline's *flight dispatcher* before departure based on the available options given the designed airspace, and is submitted to the proper stakeholders (including obviously the crew and the ANSPs) in the shape of a formal *Flight Plan* (see section 1.4). The ASM activity includes, among others, the definition of a network of routes (referred to as ATS routes), the organization of the airspace in regions and control sectors, the airspace classification (certain airspace is only flyable by aircraft fulfilling determined conditions) and other airspace constraints (some regions might be limited/restricted/prohibited for civil traffic). All this information is published in the AIP. Chapter XX will be devoted to airspace management, where all these issues will be tackled in detail.

1.3.2 Air Traffic Flow and Capacity Management (ATFM)

This is the second layer of the ATM system. It is performed during the pretactical phase, still before aircraft departure, with a time interval ranging typically from weeks up to three hours before the flight starts. Once the *Flight Plan* has been submitted by the airline according to its individual preferences, fulfilling the ASM airspace design requirements, a centralized entity named the *Network Manager* matches this Flight Plan with all the others expecting to be operating at the same time windows in the same areas in order to check whether the available system capacity is exceeded. Only a limited number of flights can be safely handled at the same time by each air traffic controller in the designated volumes of airspace under his responsibility (assigned airspace sector).

If, as a result of the analysis performed by the Network Manager, the existing demand is expected to overload a given sector (exceeding its declared capacity) then some of the flight plans are modified, resulting in reroutings and ground delays. Chapter XX will be specifically devoted to ATFM.

1.3.3 ATS Services

These services are provided at a tactical level, during the operation of the aircraft. Once the flight plan has been approved by ATFM authorities, it can be flown. As we will see in a single chapter devoted to these services, they include information, alert and control services.

Unfortunately, there are many elements that introduce uncertainty in the system (payload estimation, atmospheric conditions, measurement, piloting and modeling errors, etc) and there is always a lack of adherence between initially planified intentions (i.e. flight plan content) and the ability of the network to fulfill these intentions in real time. Thus, there must be units (named *ATS Units*) to ensure that all flights evolve safely, detecting and avoiding any potential hazard (e.g. potential conflicts) by modifying aircraft trajectories. This task, among many others, is fulfilled by the Air Traffic Control Service (ATC). ATC is executed over different volumes of airspace (enroute, approach, airport surface, etc) in different ATS Units (i.e. Area Control Centers, ACCs, and Control Towers) by different types of controllers.

1.4 Flight Plan within the ATM concept

A flight plan is an aviation term defined by ICAO as *specified information provided to air traffic services units, relative to an intended flight or portion of a flight of an aircraft*[1].

A flight plan is prepared on the ground and specified in three different manners: as a document carried by the flight crew, as a digital document to be uploaded into the Flight Management System (FMS) and as a summary plan provided to the

---

[1]This in accordance with ICAO Document 4444

Air Transit Services (ATS). It has information on intended lateral route, requested flight levels, expected speeds, overflight times, fuel consumption for each flight segment, alternate airports, and other relevant data so that the aircraft properly receives support from ATS Units in order to execute a safe operation.

Two safety critical aspects must be properly assesed:

- Fuel estimation, to ensure that the aircraft can safely reach the destination.

- Compliance with Air Traffic Control (ATC) requirements, to minimize the risk of collision.

Flight planning is the process of producing a flight plan. It requires accurate weather forecasts, not only because the most efficient route will strongly come determined by the behaviour of the atmosphere, but also because fuel consumption estimation strongly depends on the effect of head or tail wind components and static air temperature. Furthermore, ATC needs to constantly know which are the intentions of a given aircraft, and these come determined by its flight plan. An effective flight plan can reduce fuel costs, time-based costs, overflight costs and loss of revenue due to uncarried payload for overweight restrictions when fuel is overestimated, simply by efficiently modifying the intended lateral route, required altitudes, speeds, or selection of alternate airports.

From the point of view of ATS provision, the flight planning process starts when the airline declares its intention to operate a particular route and finishes when the aircraft takes off. The importance of planning can be seen from two different perspectives: on one hand, airlines have to provide an assignment to their (limited) resources; on the other ANSPs need to know the demand in advance in order to adjust their capacity.

Regulations establish that flight plans must be submitted prior to departure (except for some cases of VFR flights in non-controlled airspace, classes E, F, G) via two procedures:

- Presentation at the Airport Reservation Office (ARO office) in the airport of origin

- Presentation at the Integrated Flight Plan Service (IFPS, dependent on Eurocontrol in Europe)

IFR flight plans must be submitted at least three hours prior to the intended EOBT (Estimated Off Block Time).

2 CNS Concept

CNS stands for Communication, Navigation and Surveillance. These three services have to be deployed to provide the necessary infrastructure to ensure that the ultimate goals of air navigation are properly satisfied for each region over which

ATM services are to be provided. These three are the technical enablers that are required to support aerial operations.

Communications, navigation, and surveillance are essential technological systems for pilots (in the air) and for air traffic controllers (on the ground). They facilitate the process of determination of position, identifying and avoiding potential threats (e.g. potential conflicts with other aircraft). In order to fly from one initial point to the next one in a safe way, an aircraft must keep continuous contact with the different assigned ATS Units on earth by means of communication systems, it must use navigation systems to continuously determine its position and navigate according to its flight plan. In this whole process, ATS Units must use surveillance systems to monitor aircraft position and, getting a view of the global picture, avoid any potential loss of separation between pairs of aircraft.

The navigation pillar of the CNS concept is covered along different chapters in this book, where we will address how the aircraft can determine his position using navaids. One single chapter is also devoted to surveillance. We will here get a little bit more in detail with respect to the communication pillar of the CNS concept, and introduce briefly the two other remaining systems.

2.1 Communication

The technical means included under the term *aeronautical communication* fulfill the mission of spreading any information of interest to aircraft operations in real-time[2], typically during the pretactical and tactical phases. We are talking here about the communication channels that pilots and air traffic controllers use to issue clearances, share relevant information, coordinate actions, etc, being all these channels used to execute true actions and not just to report intentions, having nothing to do with the preplanning phase of a flight. Also, coordination between different ATS Units is required, all of them being connected to the Network Manager, and the different communication ground to ground systems have to be deployed (to transfer information regarding Flight Plan details, for example).

According to the existing channel, and the intended scope of the service, ICAO has classified the aeronautical communications in two main groups:

- Aeronautical Fixed Service (AFS): used between different terrestrial stations

- Aeronautical Mobile Service (AMS): used between terrestrial stations and aircraft (mobile stations)

2.1.1 Aeronautical Fixed Service (AFS)

As defined by ICAO in Annex 10 Vol II, the AFS is *a telecommunication service between specified fixed points provided primarily for the safety of air navigation and*

---

[2]Not to be mistaken with the Aeronautical Information Service (AIS) which contains airspace and procedure contents to be used in the planning phase of a flight.

for the regular, efficient, and economical operation of air services. This service
is typically on charge of spreading information prior to departure, related with
flight plans, meteorological information, operative state of the airspace, etc. Such
information must be transmitted to all fixed point stations (e.g. control centers)
that might consider it of interest. This information is typically generated in one
point and later distributed using specific terrestrial networks. The infrastructure
consists of voice and data networks, including:

- The Aeronautical Fixed Telecommunication Network (AFTN)

- The Common ICAO Data Interchange Network (CIDIN)

- The Air Traffic Services (ATS) Message Handling System (AMHS)

- The meteorological operational circuits, networks, and broadcast systems

- The ATS direct speech networks and circuits - typically telephone lines between different ACCs and Direct Access Intercoms (DAIs) between different
  sector controllers

- The Inter-Centre Communications (ICC)

Most data message interchange in the AFS is performed using the AFTN network.
This is a message-handling network running according to ICAO Standards, defined
as *a worldwide system of aeronautical fixed circuits provided, as part of the aeronautical fixed service, for the exchange of messages and/or digital data between
aeronautical fixed stations having the same or compatible communications characteristics.* ATFN exchanges vital information for aircraft operations such as flight
plan information, distress messages, urgency messages, flight safety messages, meteorological messages, flight regularity messages, and aeronautical administrative
messages. This network transmits messages at low speed and therefore has low
capacity meaning it is a completely outdated network, however still widely used
today. In order to create a technological upgrade to cope with the increasing volume of information, the CIDIN was conceived in the 1980s to replace the core of
the AFTN. The technology on which the CIDIN is based is referred to as *pack-
ages commutation* and it is considered as a high speed, high capacity transmission
network. Typically, most nodes which are part of the AFTN have also CIDIN
capability, and thus the CIDIN can be considered as a data transport network
which supports the AFTN. Nevertheless, the required volumes of information are
increasing more and more and CIDIN is also about to be obsolete (if not already).
The equipment and protocols upon which CIDIN is based need to be replaced
by more modern technology with new messaging requirements. To meet these
requirements ICAO has specified the ATS Message Handling System (AMHS)[3], a
standard for ground-ground communications not fully deployed yet. The AMHS is

---

[3]The standards of the AMHS can be consulted in the ICAO Doc 9880-AN/466: Manual on
Detailed Technical Specifications for the Aeronautical Telecommunication Network (ATN)

an integral part of the ATM/CNS concept, and it is associated to the Aeronautical Telecommunication Network (ATN) environment[4].

The goal of ATN is to be the *aeronautical internet*, a worldwide telecommunications network that should allow any aeronautical actor (ATS units, airlines, private aircraft, meteorological services, airport services, etc) exchange information in a safe way (such as ATC instructions, meteo messages, flight parameters, position information, etc.) under standard message formats and standard communication protocols. The European AMHS makes use of a TCP/IP network infrastructure, in line with the recent evolution of the ATN concept for ground communications. In addition to being the replacement for AFTN/CIDIN technology, the AMHS also provides increased functionality, in support of more message exchanges than those traditionally conveyed by the AFTN and/or CIDIN. This includes, for example, the capability to exchange binary data messages or to secure message exchanges by authentication mechanisms.

2.1.2 Aeronautical Mobile Service

On the other hand, the aeronautical mobile service includes all technical means required to support the communications between aircraft and ATS units. These communications are pilot-controller. As defined by ICAO Annex 10 Vol II the aeronautical mobile service *is a mobile service between aeronautical stations and aircraft stations, or between aircraft stations, in which survival craft stations may participate.*

The ultimate goal of this service is to allow communications between pilots and controllers. In particular, within the boundaries of one control sector, the controller must be able to communicate with all aircraft inside the sector using only one of these radio channels, that is, each sector has a unique assigned frequency that must be monitored by all aircraft within the sector, listening not only to the instructions of the air traffic controller but also to all the other aircraft in their vicinity. Therefore, the number and dimension of sectors conditions the location of the communication antennas, to provide enough and reliable signal coverage. The frequency assigned to each sector establishes a double direction channel: pilot-controller; controller-pilot. These channels are the fundamental instruments used when executing the functions of information, alert and control.

There are two types of pilot-controller communication:

1. **Controller-pilot voice communications** - These services are provided wirelessly, using radio channels. In the case of aeronautical communications, the VHF (Very High Frequency) and HF (High Frequency) bands are used. VHF channels have their range limited to line-of-sight while HF signals can bounce against the ionosphere, extending their range and being typically used as part of the CNS concept when applied over big depopulated regions

---

[4]The standards of the ATN can be consulted in the ICAO DOC 9705-AN/956: Manual of Technical Provisions for the ATN.

such as the North Atlantic or Siberia. The channels in HF are only used for long-distance communications, when it is impossible to establish communication using VHF. Voice quality is poorer in HF than in VHF. VHF radio communications for civil aviation operate in the frequency range extending from 118MHz to 137MHz. HF radio communications use practically the whole HF spectrum (3 MHz to 30MHz), depending on times of the day, seasonal variations, solar activity, etc. All aircraft are required to constantly monitor in their standby equipment the VHF 121.5MHz frequency, which is used in case of emergency.

2. **Controller-pilot datalink communications (CPDLC)** - A mean of communication between controllers and pilots that uses datalink procedures and where messages are presented as text instead of voice. Messages can be transmitted using VHF or satellite bands. When either the pilot or the controller wants to establish contact, a message containing the request/instruction is sent. Figure [15.5] illustrates a pilot CPDLC interface.

2.2 Navigation

Navigation systems allow aircraft to know their positions at any time. It is important to distinguish between the systems that assist pilots to steer their aircraft (navigational aids), and the techniques that pilots use to navigate. Navigational aids constitute infrastructures capable of providing pilots with all the required information in terms of position and guidance. On the other hand, the navigation techniques refer to the way in which pilots use these data about position to navigate.

The navigational aids systems can be classified in two main groups:

- Autonomous systems: Those that make use only of the means available in the aircraft to obtain information about its own position

- Non-autonomous systems: Those external systems that provide the aircraft with the information about its position

There are two chapters in this book devoted exclusively to non-autonomous systems of navigation (ground and space-based navaids), so for now we will complement this introduction of the ATM/CNS concept extending a bit the information about autonomous systems for the determination of position.

2.2.1 Autonomous systems

Using only autonomous navigation systems, the most advanced navigation technique to be used is named *dead reckoning* (we briefly discussed about this technique in our first chapter). The dead reckoning technique is based on predicting the future position of the aircraft as a function of the current position, expected ground speed and track. Obviously, an initial position of the aircraft must be known. In order to determine this initial position, different means can be utilized, for instance, observing a point near the aircraft whose position is known (very rudimentary), the observation of celestial bodies (also rudimentary), or the use of the so-called *autonomous systems*, which are able to determine the ground speed and track of the aircraft.

The mainly used autonomous system is the Inertial Navigation System (INS). An Inertial Navigation System (INS) includes at least a computer and a platform or module containing accelerometers, gyroscopes or other motion-sensing devices. The later one is referred to as Inertial Measurement Unit (IMU). The computer performs all navigation calculations. The INS is initially provided with its position and velocity from another source (a human operator, a GPS satellite receiver, etc) and thereafter computes its own updated position and velocity by integrating the information received from the motion sensors. Figure [4.7] illustrates how an INS works schematically. The advantage of an INS is that it does not require any external reference in order to determine the aircraft´s position, orientation or ground speed once it has been initialized. However, the accuracy of this instrument is limited, specially after long times of integration.

There are two fundamental inertial navigation systems:

- Stable platform systems (aligned with the global reference frame)

- Strap-down systems (aligned with the body frame)

Gyroscopes measure the angular velocity of the aircraft in the inertial reference frame (for instance, the earth-based reference frame). By using the original orientation of the aircraft in the inertial reference frame as the initial condition and integrating the angular velocity, the aircraft's orientation (attitude) can be known.

All inertial navigation systems suffer from integration drift: small errors in the measurement of acceleration and angular velocity are integrated into progressively larger errors in velocity, which are compounded into still greater errors in position. Since the new position is calculated from the previous calculated position and the measured acceleration and angular velocity, these errors are cumulative and increase at a rate roughly proportional to the time since when the initial position was input. Therefore the position must be periodically corrected as input from some other type of navigation system. The inaccuracy of a good-quality navigational system is normally less than 0.6 nautical miles per hour in position and on the order of tenths of a degree per hour in orientation. Inertial navigation is usually supplemented with other navigation systems (typically non-autonomous systems), providing a higher degree of accuracy. The idea is that the position (in general, the state of the aircraft) is measured with some sensor (e.g. GPS) and then, using filtering techniques (Kalman filtering, for instance), one can estimate the position based on a weighted sum of both measured and predicted position (the one resulting from inertial navigation). The weighting factors are related to the magnitude of the errors in both measured and predicted position. By properly combining both sources of information, the errors in position and ground speed are nearly stable over time.

2.3 Surveillance

The technical means included under the term *aeronautical surveillance* fulfill the mission of providing real-time information containing the position of a set of aircraft within a given airspace sector to ATC units (i.e. air traffic controllers), with the aim of ensuring safety by properly separating them, executing actions to avoid any potential loss of separation. As we will analyze in one separate chapter, there are different surveillance methods. It has been traditionally carried out using radar signals. However, current surveillance satellites, together with Automatic Dependent Surveillance Broadcast (ADSB) technologies will be replacing (sooner, rather than later) radar as the primary surveillance method to control aircraft worldwide.

There are also airborne systems that fulfill a surveillance function. This is the case of the Traffic Collision Avoidance System or the Traffic alert and Collision Avoidance System (both abbreviated as TCAS), which act as automatic advisory back-up systems when imminent threats (intruders) get too close to a given aircraft (i.e. once the human-based ATC safety layer has failed), trying to prevent a mid-air collision.

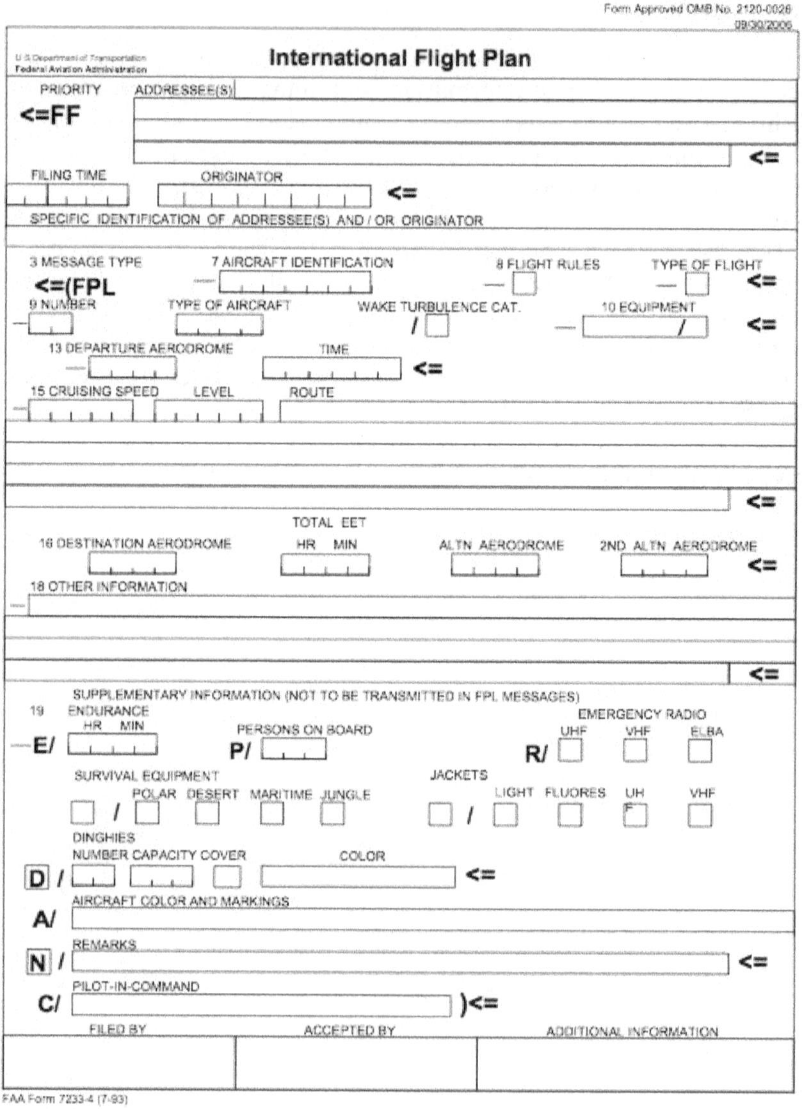
Form Approved OMB No. 2120-0026
09/30/2006

U.S. Department of Transportation
Federal Aviation Administration

International Flight Plan

PRIORITY ADDRESSEE(S)
<=FF
<=
FILING TIME ORIGINATOR <=
SPECIFIC IDENTIFICATION OF ADDRESSEE(S) AND / OR ORIGINATOR

3 MESSAGE TYPE <=(FPL
7 AIRCRAFT IDENTIFICATION
8 FLIGHT RULES
TYPE OF FLIGHT <=
9 NUMBER
TYPE OF AIRCRAFT
WAKE TURBULENCE CAT. /
10 EQUIPMENT / <=
13 DEPARTURE AERODROME
TIME <=
15 CRUISING SPEED LEVEL ROUTE
<=
16 DESTINATION AERODROME
TOTAL EET HR MIN
ALTN AERODROME
2ND ALTN AERODROME <=
18 OTHER INFORMATION
<=

SUPPLEMENTARY INFORMATION (NOT TO BE TRANSMITTED IN FPL MESSAGES)
19 ENDURANCE HR MIN
E/
PERSONS ON BOARD P/
EMERGENCY RADIO UHF VHF ELBA R/
SURVIVAL EQUIPMENT POLAR DESERT MARITIME JUNGLE /
JACKETS LIGHT FLUORES UHF VHF /
DINGHIES NUMBER CAPACITY COVER COLOR
D / <=
AIRCRAFT COLOR AND MARKINGS
A/
REMARKS
N / <=
PILOT-IN-COMMAND
C/)<=
FILED BY ACCEPTED BY ADDITIONAL INFORMATION

FAA Form 7233-4 (7-93)

Figure 4.4: Flight Plan FAA International Form © FAA

Figure 4.5: VHF vs HF range (left) & basic VHF radio selecting equipment, with emergency as standby frequency (right) © F1jmm / Wikimedia Commons / CC-BY-SA-3.0

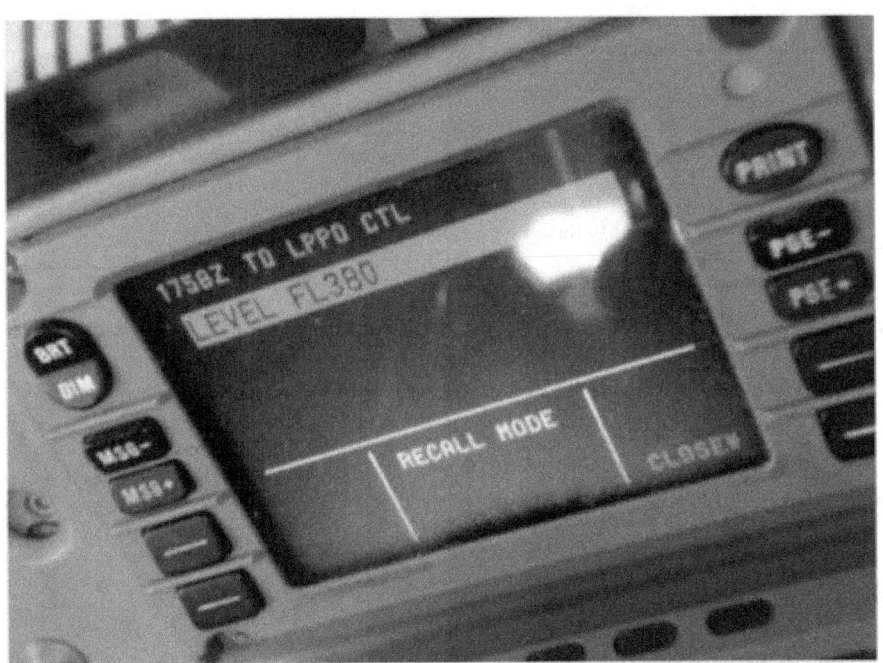

Figure 4.6: Datalink control and display unit (DCDU) on an Airbus A330 ©SempreVolando / Wikimedia Commons / CC-BY-3.0

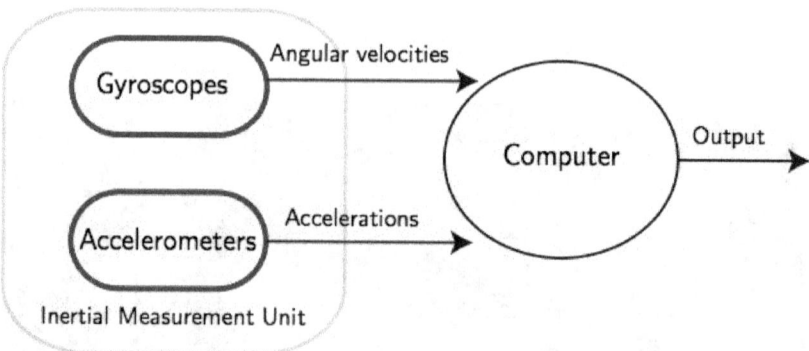

Figure 4.7: Scheme of an Inertial Navigation System (INS). The output refers to position, attitude, and velocity.

Chapter 5. Surveillance

1 Introduction

Talking about surveillance means talking about one of the three basic pillars that support the CNS[1] concept. Surveillance consists of determining the position (lateral and vertical) of any vehicle or object within a given sector of airspace. A proper and efficient surveillance is thus absolutely basic to understand what and how are "things" flying within a given piece of airspace. Surveilling means watching closely what is happening (having enough data to depict the actual and current position of all traffic), in terms of lateral and vertical position as well as basic kinematic properties (how fast and in which direction are aircraft moving). Surveillance systems together with flight plan information databases let air traffic controllers predict what are the intentions of different aircraft (where they will fly next).

In order to provide the three basic ATS Services[2], a reliable surveillance system is required (e.g. how could we possibly inform a pilot about the position of another aircraft if the information concerning their positions were unreliable?). Surveillance systems are characterized by the levels of reliability, tolerance and completeness in the data they retrieve (e.g. are we able to capture information about the position of all vehicles within a given airspace sector? do we require their collaboration and thus those who do not collaborate might be not reporting their presence? how accurate do we know about lateral and vertical positions of the different aircraft?). Needless is to say that surveillance requirements may be different depending on the application or use that is going to be provided to the users of the airspace, being for example military requirements (to prevent non-authorized airspace incursions) different from civil ATC requirements. Also, the refreshment rate of the information retrieved determines operational procedures (e.g. a more accurate and refreshed system will decrease the uncertainty associated to the distance between a pair of aircraft, enabling thus shorter minimum separation requirements).

---

[1]Communication, Navigation and Surveillance
[2]Alert, Information and Control, as we will see in a future Chapter

2 Surveillance methods

Although radars are the most extended and known surveillance systems, there are other methods worth mentioning:

- **Observation** - specially in the vicinity of aerodromes, still today the most reliable surveillance method is the use of human eyes assisted by binoculars. Used to determine the position of aircraft in the different legs of a visual pattern, to determine whether the runway is clear or not, the presence of animals, etc. Runway allocation (basic control service task for a tower air traffic controller) is still today done taking into account a lot of information retrieved by the controller from his visual contact with the aircraft (LVP are an example of increased minimum required separation because of a lack of this surveillance source). However, observation can sometimes lead to misleading judgement, specially when the objects are far away from the observer and there is a wrong perception of their position.

- **Noise detection** - at the very first times of aviation, this was a technique to detect airspace incursions by enemy war planes. Figure [5.1] shows an example of this technology, that turned completely obsolete after the appearance of radars. However, still today automated noise detection tools are used in the vicinity of aerodromes to monitor noise levels and the associated environmental impact.

Figure 5.1: World War I noise detection technique © skylighters

- **Radio reports** - pilot-based reports through the radio let air traffic controllers take notes and write down which is the position of aircraft according to pilots, their evolving flight levels and their time estimates to the different flight points. Still today these techniques are used for wide oceanic areas where no radar is available. *Procedural control* is the name of the control technique used when separating aircraft from each other based on radio reports, using obviously larger safety margins than when using accurate radars, and resulting thus more inefficient. In case of radar failure, some ATS units may use these techniques as back up procedures. Figure [5.2] shows different flight progress strips used as a back up surveillance method where information transmitted through radio reports is collected (colors and positions of the different strips on the table have different meanings). Procedural air traffic control is a complete science by itself, with high complexity, intense workload and radio communication required for a safe and efficient management of the airspace.

Figure 5.2: Flight progress strips used for aircraft tracking and radio-transmitted surveillance

2.1 PSR and SSR

During WWII, first radars appeared to detect airspace incursions by the enemy. Radar stands for "Radio Detection and Ranging", and consists of a high frequency radio transmitter broadcasting high energy short-length radio waves through a rotating antenna, being a small proportion of the energy reflected by the surface/structure of aircraft, travelling back to the radar receiver. First developed radars tried to detect the presence of obstacles through radio reflection effects, measuring the time it takes the radio wave from leaving the transmitter to hitting the receiver on the way back (and thus computing an estimation for the distance at which the obstacle is found), azimut of the antenna providing the bearing of the aircraft from the ground station. Figure [5.3] depicts the basic functioning of a PSR radar.

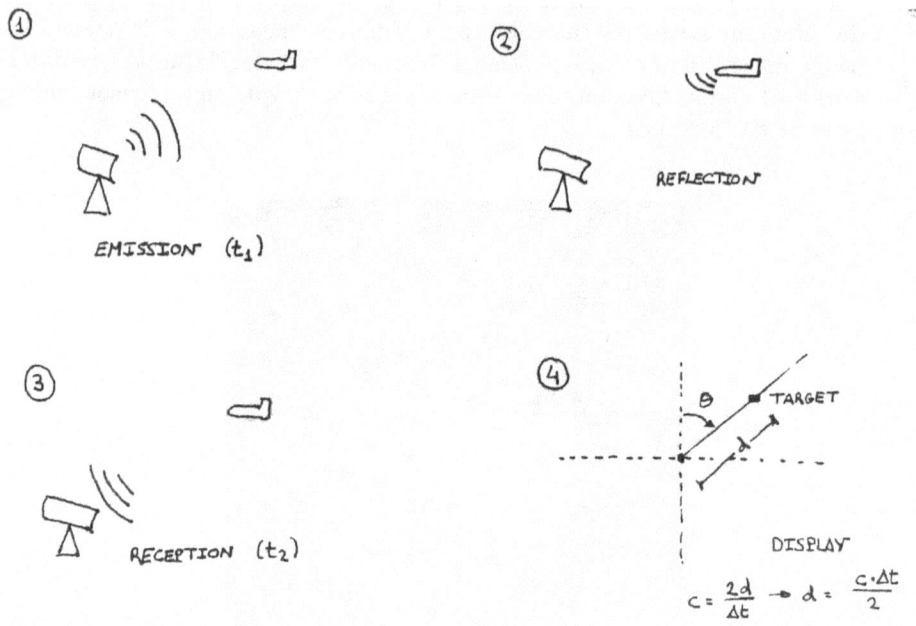

Figure 5.3: PSR basics

These initially developed radars work as PSRs (Primary Surveillance Radars) and are used to detect vehicles or objects through the use of echo principles (they do not need any collaboration at all from the object itself). PSR radars send radio signals for given azimut angles and for all elevation angles, being thus unable to detect the altitude of the targets, only militar advanced radars (3D) can detect the altitude of non-collaborative targets. The important limitations of PSR radars are:

- Important amounts of power have to be radiated to make sure part of the energy can echo

- Echo returns easily fade away, specially under adverse weather conditions (e.g. clouds or rain)

- Correlation[3] is not easily done, requiring extra workload such as extra vec-

---

[3]Correlation is the process of identifying a target with a known aircraft callsign and its associated flight plan (e.g. knowing that the dot displayed on the screen, detected by a radar, truly corresponds to IBE3485)

toring or confirmation of position report through the radio. Old school air traffic controllers used to instruct airplanes to turn in order to perform correlation and identify that the aircraft turning truly was the aircraft they thought it was.

- Coverage is limited to the line of sight (areas behind mountains or big obstacles remain in the shadow for the radar)

- Only provides lateral information regarding the position of the target (no altitude information)

- PSR radars often see too much for air traffic control purposes, since all stationary objects do also reflect the emitted signal. Filters are then necessary to remove from the presentation those objects that are not relevant for the controller. This problem is named Ground Clutter and is normally solved using a MTI (Moving Target Indicator).

Some of these problems we have numbered regarding PSRs are solved through the use of SSRs (Secondary Surveillance Radars), that requiring collaboration from the target, can get a lot more information. A SSR sends signals that contain different questions (as we will see in section 2.2) that a system onboard will detect and answer (named *transponder* or *squawk*). We can not forget that a SSR always relies on aircraft systems and requires transponder collaboration. For cases in which collaboration can not be expected (as happened during the attacks of September the 11th in NYC) transponders can be switched off and targets on the screen just disappear as no answer is received by ground SSR antennas. Reinforced signals, where a combination of both PSR and SSR signals are merged, is then preferable for those cases in which aircraft transponders can fail.

2.2 SSR Transponder modes

Depending on the amount of information the transponder of the aircraft sends back to the SSR antenna, we can define the following transponder modes[4].

- **Alpha Mode:** tries to solve correlation. An alpha mode transponder sends back a code of four numbers in octal (ranging from 0 to 7) that are selected by the pilot in the cockpit (see Figure [5.4]). Air traffic control systems assign a single transponder code to each aircraft (there are 4096 available codes), these codes are displayed on the radar screen, letting air traffic controllers identify aircraft. Modern correlation systems assign transponder codes to each flight plan, typically displaying directly the callsign associated to a given transponder code, as can be seen on Figure [5.5]. It is very important to preserve the one-to-one relationship, not assigning the same transponder code to two different aircraft. Current traffic levels impose the existence of

---

[4]The most complex modes include the information contained in the simplests

standards to avoid transponder coincidence and reassignment of transponder codes during the flight[5]. Certain transponder codes are used for contingency or standard situations (e.g. 7600 for the case of communication failure, 7700 for emergency situations, 2000 in Europe for aircraft flying within controlled airspace without assigned transponder code or 7000 typically for VFR traffic flying in uncontrolled airspace without assigned squawk code).

Figure 5.4: Code selection display on the cockpit © Ahunt / Wikimedia / Public Domain

 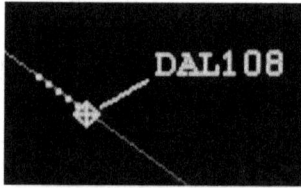

Figure 5.5: Manual (alpha code displayed) and automatic (transponder code automatically assigned to flight plan callsign) correlation

- **Charlie Mode:** answers back with altitude information. Retrieved from the static ports of the aircraft, the transponder sends these pressure values back to the SSR antenna. Altitude information is then displayed attached to the callsign of the aircraft on the screen. It must be strongly understood that this altitude information (pressure information to be more exact) has its source in the sensors of the aircraft, and is not retrieved by measuring any timing in the reflection of the echo signal[6]. Radar displays and software on the ground take into account local QNH as well as relative position of the aircraft with respect to the transition altitude and transition level, displaying the right altitude (or FL) within each of the different layers (not depending

---

[5]For more information about this, see ICAO and Eurocontrol documentation about the Originating Region Code Assignment Method (ORCAM)

[6]This misunderstanding partially led to the Aeroperu 603 accident in 1996, whose CVR can be checked in youtube.com/watch?v=oZIjh7d7JmQ (showing both, pilots and ATCOs very poor understanding of the source of the information displayed on the radar screen)

thus on the Kollsman window selection of the pilot), according always to the received pressure information from the transponder of the aircraft.

Figure 5.6: Charlie mode information, displaying aircraft Flight Level (for this case, aircraft flying at FL360, or 36000ft over the 1013 mb isobar)

- **Sierra Mode**: is the most complete transponder mode and is still today not used by most ANSPs[7]. Traditional radars can easily measure the Ground Speed ($\frac{\Delta s}{\Delta t}$) and track of an aircraft (given two consecutive positions, one can measure the angle this line forms with the True or Magnetic North, retrieving aircraft´s track); however, even though these magnitudes drive air navigation performance, they are not easily controllable by the pilot (that, instead, can easily select aircraft´s Indicated Airspeed and Magnetic Heading). As we have already seen in Chapter 3, GS differs from TAS and CAS depending on aircraft altitude and wind conditions. Sierra mode gives information regarding IAS, Magnetic Heading, Mach Number, Selected (FCU or MCP) altitude by the pilot, etc. Interrogations are sent using 1030MHz carriers while the answers come back in 1090MHz (same frequency as ADSB, see section 2.4).

2.3 Computerized radar systems

So far we have talked about the different information one can retrieve using surveillance techniques and the required equipment onboard. As important as the kind and quality of the information we retrieve, is the way we present this information to the air traffic controller. Different companies have developed software and radar working positions that integrate the machine/system with which air traffic controllers have to operate. In Spain, for example, Indra developed SACTA (Sistema para la Automatización del Control de Tránsito Aéreo).

Computerized radar systems work with flight plan databases making automatic correlation (displaying callsigns instead of raw transponder codes) and integrate information about the status of the network (available radars, runways in use,

---

[7]A video showing how Sierra Mode has been implemented in MUAC (Maastricht Upper Area Control Center) can be found in https://www.youtube.com/watch?v=9cQKPiGvD5Y

weather information, etc). As an example of the information that is usually displayed, we will comment on Figure [5.7].

Figure 5.7: IBE04TU departing from LEBB on his way to LEMD

- IBE04TU is the callsign of the aircraft. This information is not sent by the aircraft's transponder (that instead, sends a transponder code). The correlation process is supported by ground flight plan databases.

- 102↑ is called AFL (Actual Flight Level) and reflects Charlie mode information about the current static pressure sent from the aircraft. The upper arrow indicates the aircraft is climbing (previous AFL was lower than 102).

- 140 is called CFL (Cleared Flight Level) and is manually introduced by the Air Traffic Control Operator (ATCO). It is the altitude (or level) to which the aircraft has been cleared. Although it does not appear on Figure [5.7], the system may also inform the ATCO about the flight plan RFL (Requested Flight Level), that may evolve in time as the aircraft burns fuel (stepped climb). However, flight plan RFL and tactical RFL do often differ (adding lack of adherence to ATFM expectations) because typically flight plans are repetitive (same flight plan for a set of days, or months) while conditions (aircraft weights, atmosphere temperatures, winds, etc) are different each day.

- APPBIL shows the ATS Unit currently controlling the aircraft, in this case, Bilbao Approach.

- 30 shows the Ground Speed of the aircraft (300 knots) computed by the ground systems, based on the evolution of the target in time.

- M gives information about the wake turbulence category of the aircraft (a flight plan was made in which the pilot specified the aircraft type, an Airbus 320, which is a Medium wake category aircraft).

- MD shows the aircraft flight plan destination aerodrome (in this case, LEMD, Madrid).

- 32L is the runway assigned to this aircraft for landing

- 07 is part of the AMAN information (Arrival Manager) and reflects the queue number assigned to this aircraft in the arrival sequence (that is manually modified depending on traffic situation by Madrid Approach ATCOs).

A deep study concerning the presentation of information on the screen has to be made prior to its implementation. Human factors and ergonomy of the presentation play a key role in the way the information is understood. We should not think that the more information we present, the better we are depicting a traffic situation, since we might be overwhelming the controller. Sometimes, the simpler the better. Most systems today are capable of providing a whole set of complete information by request from the controller, but they do not present this information by default (the controller can select which information he wants to see), an example of this can be traffic filters for certain flight levels.

Computerized radar systems do normally incorporate conflict algorithms that pop up when a loss of separation is expected or is happening. However, its calibration is not always perfect and there are many false positives (making the ATCO less sensitive to the alert). A case of this, in which a false echo appears, popping up a false loss of separation is depicted on Figure [5.8].

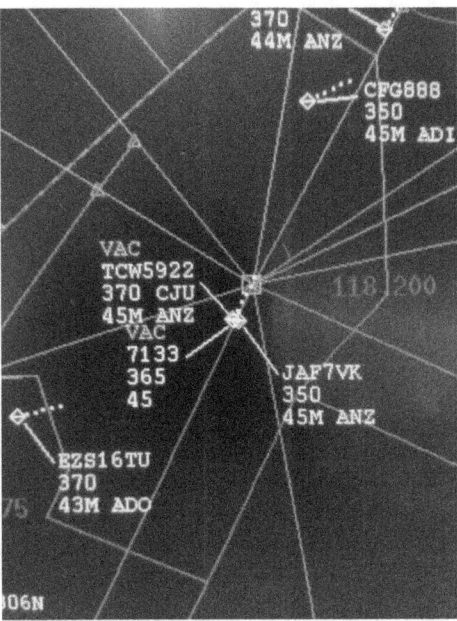

Figure 5.8: False positive alert triggered because of false target appearance (squawking 7133 at FL365 below TCW5922 at FL370)

2.4   New surveillance systems. ADSB & ADSC

New surveillance techniques are constantly under research, looking for a safer and cheaper management of the airspace. Multilateration (MLAT) starts to be broadly implemented in airport environments (substituting surface radars). MLAT works by measuring, at different ground facilities whose position is very well known, the differences in time of a signal emitted by the aircraft, permitting the determination of aircraft´s position. However, here we will shortly talk about ADSB and ADSC, that despite of having similar names, have almost nothing to do with each other.

ADSB stands for Automatic Dependent Surveillance Broadcast. An ADSB surveillance network consists of a set of receiving antennas that receive messages sent from aircraft squitters in the 1090MHz frequency. Aircraft squitters need no interrogation and are constantly emitting different messages (in hexadecimal code) with different flight parameters (typically radiating position, altitude, vertical speed, ground speed, track, callsign, squawk code, etc). All flight parameters are obtained through the different sensors of the aircraft and the information obtained is thus 100% *dependent*. If the information retrieved by the aircraft is wrong, so will it be radiated. Today, there are still important constraints in terms of the reliability and completeness of ADSB data. It is mostly used for online applications and passenger information[8].

On the other hand, ADSC stands for Automatic Dependent Surveillance Contract, and consists of a system through which aircraft can radiate their position (and other details) using a satellite link (which acts later as a relay to other ground stations). ADSC is used when PSR or SSR are not available (e.g. wide oceanic areas), being the cost of the message an important restriction. Today, ADSC together with CPDLC (Controller Pilot Datalink Communication), INS (Inertial Navigation Systems) and GNSS (Global Navigation Satellite System), represent the way the CNS (Communication, Navigation and Surveillance) concept is developed in oceanic navigation.

3   TCAS and onboard radars

TCAS stands for Traffic Collision Avoidance System and consists of an interaction system installed onboard that lets transponders from different aircraft "talk" to each other, reporting the relative position of other aircraft in the vicinity to the pilot. It is independent from any ground infrastructure. Traffic is typically represented on the Navigation Display (see Figure [5.9]) or on the variometer (older avionics). Each diamond represents an aircraft near the observer, and the displayed number indicates its relative vertical position with respect to the observer (in hundreds of feet). TCAS works taking into account the relative speeds with which different aircraft come closer or farther from the observer, alerting and even

---

[8]Webpages such as fr24, planefinder or flightaware, have created communities of receivers that share data

advicing in case of potential collision. Depending on the degree of the loss of separation, different colours are used when representing the affecting traffic on the navigation display[9].

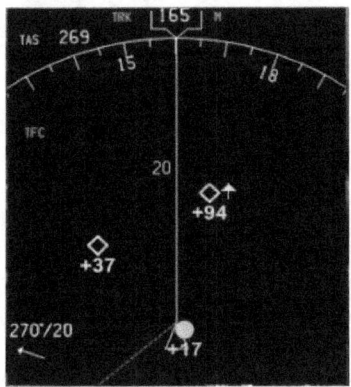 

Figure 5.9: TCAS representation of several aircraft on a ND (left) and on a variometer (right)

TCAS outputs two different types of messages:

- **Traffic Advisory (TA)**. When aircraft in the vicinity come too close to the observer, developing a potential risk of collision if the situation deteriorates, also alerting the crew that a RA may be triggered. Diamonds on the display would get full white or yellow, and the speaker would warn "TRAFFIC" loudly.

- **Resolution Advisory (RA)**. These messages require the immediate intervention of the crew, disconnecting the autopilot (if necessary) and following the RA instruction (climbing or descending). RAs always imply a vertical avoidance maneuver. RA messages are triggered when there is real danger of collision. Transponders "talk" to each other agreeing a common strategy to avoid the collision (instructing one of the aircraft to climb and the other to descend)[10]. Variometer areas are illuminated in a red-green distribution, instructing the pilot which rate of climb or descend has to be acquired. The speaker warns loudly with the action that has to be followed (i.e. "CLIMB, CLIMB" or "DESCEND, DESCEND").

It must be stated that TCAS is the last safety net before a mid-air collision, that can only be triggered if both aircraft have it operative and their transponders are

---

[9]A video in https://www.youtube.com/watch?v=zk7dl3NStIo shows how information presented by TCAS matches real time crossing with other traffic

[10]For a simulation on how RAs are displayed on a variometer, check https://www.youtube.com/watch?v=z-6zF9PEtdU

switched on[11]. In case TCAS RA is triggered, crew has to inform ATC (that should stop providing instructions) and never disobey TCAS[12]. However, even though TCAS is essential for the safety of current air operations, its level of maturity can not lead us (yet) to completely rely on it to maintain aircraft safely spaced, there are still many cases in which TCAS is falsely activated (false positives, specially when approaching a cleared flight level with excessive vertical rate, being another aircraft at the immediate superior or inferior flight level in the vicinity) and cases in which, due to the geometry of the conflict or the presence of third parties (other aircraft affecting the geometry of the crossing), the proposed solution is non-existent or invalid. Air navigation professionals should not forget about their duties assumming TCAS is perfect (because it is not, equally to other safety barriers, it has its own limitations). Article 2.4.2 of ICAO Annex 11 clearly states that *the carriage of airborne collision avoidance systems (ACAS) by aircraft in a given area shall not be a factor in determining the need for air traffic services in that area.*

Besides TCAS, aircraft have onboard radars (situated in the nose of the aircraft, inside the radome) that can surveil for the presence of adverse weather phenomena (CBs) or obstacles (terrain), specially useful when aircraft are flying during the night or in IMC conditions[13]. Lack of proper understanding or malfunctioning of this radar can lead to entering areas of severe turbulence and ice formation[14].

---

[11]This was not the case in the mid air collision between a private jet and a B738 over Brazil in 2006. More details can be found in http://www.skybrary.aero/bookshelf/books/546.pdf

[12]This was sadly not the case in the Überlingen accident in 2002, more information can be found in http://www.skybrary.aero/index.php/T154_/_B752,_en-route,_Uberlingen_Germany,_2002_(LOS_HF)

[13]In https://www.youtube.com/watch?v=w8IYyFmJcF0 one can check a Honeywell video with nice shots of the presentation of weather information on the Navigation Display

[14]An example of this is an Air France A330 incident that happened in February 2012 and that can be consulted in http://www.bea.aero/docspa/2012/f-cg120227.en/pdf/f-cg120227.en.pdf

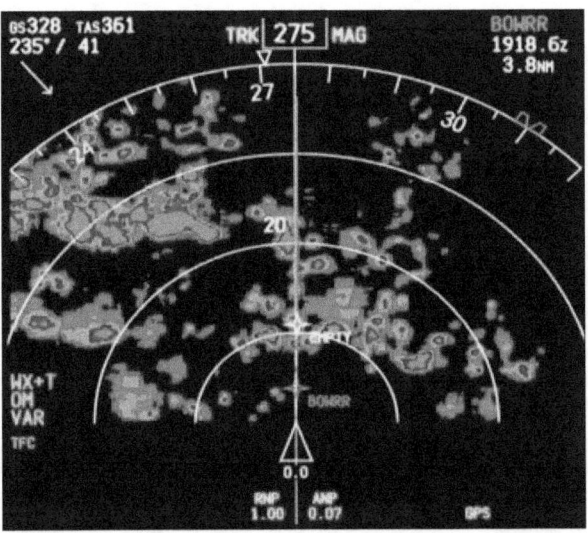

Figure 5.10: Weather information on a Navigation Display

Chapter 6. Ground Navaids

The determination of an aircraft´s position with respect to the desired route is key in Air Navigation. In doing so, aircraft are supported by autonomous or non-autonomous equipments.

Aircraft may autonomously estimate their position (typically in global geographic coordinates) through the use of Inertial Navigation Systems (INS). These consist of a set of accelerometers and gyroscopes that integer the trajectory of an aircraft measuring accelerations and time lapse, provided an initial position is known.

However, even though the use of these integering functions gives smooth trajectory outputs, errors increase as time goes by, degrading the navigation performance.

The use of external navigation aids (commonly referred to as *navaids*) lets aircraft determine their position with respect to a given reference framework. These navaids may be ground or space-based. The observer will typically get his relative position with respect to the navaid. If he wanted to obtain his global position, the exact position of the navaid on Earth should be previously known by the navigator. Their use is based on the exchange of radiofrequency signals that transmit messages to and from the aircraft. The use of different radiofrequency signal properties has enabled the development of techniques that support the determination of the relative position of the receiver with respect to the navaid.

Low Frequency Radiorange (LFR) was used during the 30s and 40s, but its strong performance limitations (it had only four legs) and the development of more advanced ground navaids gave way to the ones described along the following lines, which, despite of having been in use for decades, remain still today being deployed and used all around the world.

1 Non-Directional Beacon

A Non-Directional Beacon (NDB) consists of:

- A ground transmitter placed at a known location which radiates a radiofrequency signal that travels at the speed of light and whose frequency ranges in the hundreds of kHz. Each NDB is named with a one, two or three letter code that is also transmitted in morse together with the NDB signal to facilitate identification.

- A couple of antennas at the receiver, a decoder and an instrument named ADF (Automatic Direction Finder). This pair of antennas are placed aligned with the axis of the aircraft at a distance $d \ll \lambda$ from each other. The broadcasted signal will hit each of the antennas at a different time, with a different phase. By measuring the difference in phase of the radiated signal between each of the antennas, one can determine the direction from where the signal is coming (this technique is named radiogoniometry).

The radiated wave can be assumed to have an expression of the electric field such as $\overrightarrow{E}(r,t) = \frac{\overrightarrow{E_0}}{4\pi r} \cos[\omega\tau] = \frac{\overrightarrow{E_0}}{4\pi r} \cos\left[\omega\left(t - \frac{r}{c}\right)\right]$ where $\overrightarrow{E_0}$ provides the direction and amplitude of the electric field, r is the distance to the transmitter and τ is the time when the signal left the transmitter (which is previous to instant t of reception), being c the speed at which the wave travels.

The receiver is assumed to be far enough from the transmitter so as to consider that the straight lines between the antennas and the transmitter are parallel to each other. At a given instant t, the two antennas placed on the aircraft will receive signals (see Figure [6.1]):

$$\overrightarrow{E_1} = \frac{\overrightarrow{E_0}}{4\pi r_1} \cos\left[\omega\left(t - \frac{r_1}{c}\right)\right]$$

$$\overrightarrow{E_2} = \frac{\overrightarrow{E_0}}{4\pi r_2} \cos\left[\omega\left(t - \frac{r_2}{c}\right)\right]$$

being thus the phase difference

$$\varphi_2 - \varphi_1 = \frac{\omega}{c}(r_1 - r_2)$$

and given the geometry of the antennas, we know that

$$r_2 - r_1 = d \cdot \cos\theta$$

so

$$\theta = \arccos\left[\frac{(\varphi_1 - \varphi_2) \cdot c}{d \cdot \omega}\right]$$

A needle on a Radio Magnetic Indicator (RMI) or Horizontal Situation Indicator (HSI) will point towards the NDB ground station, as can be seen on Figure [6.2].

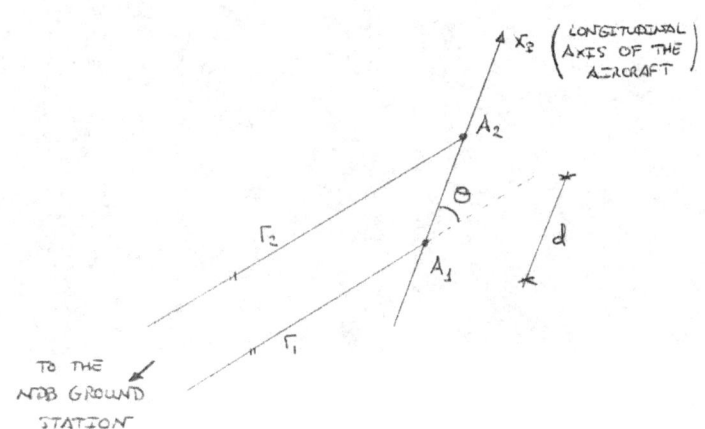

Figure 6.1: NDB phase disagreement based on geometric disposition

With an ADF one can easily navigate to or from the ground station. Also, tracking constant-distance arcs with respect to the NDB is possible (keeping the needle of the ADF perpendicular to the flight track).

A single NDB station does not provide complete position awareness to the pilot. The ADF will just point towards the ground station, not indicating how far away this station is. For a complete determination of position, three possible solutions are proposed:

- Using a DME equipment associated to the NDB station (see section 4).

- Using a second NDB station to which bearing can also be determined. The intersection between the bearings to both ground NDBs can be computed to position the aircraft.

- If the ground speed of the aircraft is known, an arc (perpendicular to the NDB) can be flown. Measuring the number of crossed radials in a given time period, one can determine the distance to the station.

2 VHF Omnidirectional Range

The VOR is probably the most popular navaid. It is used along different flight phases, including non-precision approach environments. The onboard instrument provides lateral guidance with respect to a VOR ground station, from where radiofrequency signals are sent. A VOR defines 360 equally spaced radials, taking the Magnetic North as a reference.

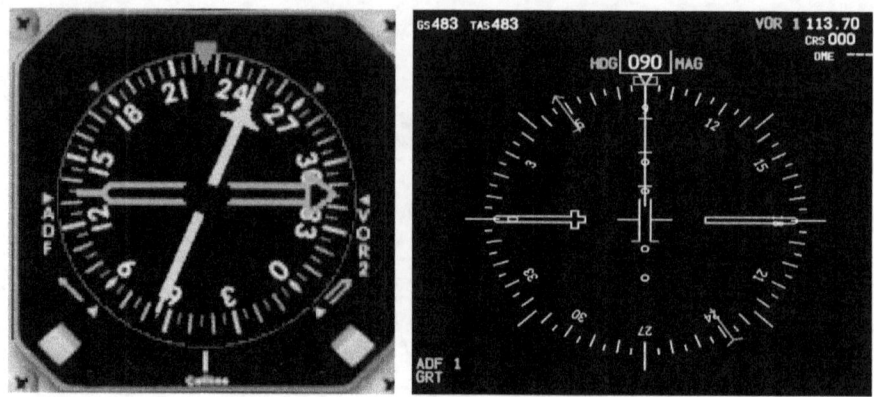

Figure 6.2: ADF Instrument on a RMI (left) and HSI (right)

The basic principles of the VOR are simple[1]. The ground station broadcasts two different signals which are coherent (both are 30 Hz signals). First an omnidirectional reference signal, through the use of a set of carriers and subcarriers that combine amplitude and frequency modulations. Second, a directional signal sent in each angle direction with a difference in phase with respect to the omnidirectional signal equal to the angle formed between the direction of transmission and a reference direction (typically the Magnetic North). This disphase is made by delaying the transmission of the directional signal with respect to the reference signal an amount equal to the value of the radial. Figure [6.3] shows both the reference and directional signals for different radials of the VOR.

Conventional VOR signals easily reflect against nearby terrain or water, overlapping other broadcasted signals and contaminating them. For this reason, they were replaced by much more accurate Doppler VORs. Doppler VORs modulate the reference signal in a different manner (avoiding the use of frequency modulated subcarriers) and make use of a couple of rotating antennas that broadcast non-modulated 30Hz radiosignals which, because of Doppler effects, make the received directional signal have different phases along each space direction (similarly to CVORs). The onboard equipment is equal for both Conventional and Doppler VORs.

The equipment onboard, after a set of filtering processes, makes a phase comparison between the directional and the reference signal ($\phi = \varphi_D - \varphi_R$), determining the radial with respect to the ground station. This phase disagreement is presented to the pilot in the VOR instrument. This instrument can be used not only to show the radial in which the aircraft is, but also to permit the selection of other intended radials and display which is the relative situation of these radials with respect to the position of the aircraft. The CDI (Course Deviation Indicator)

---

[1]An interesting interactive explanation on how a VOR works can be found in digitalflightinstructor.com/how-a-vor-works/

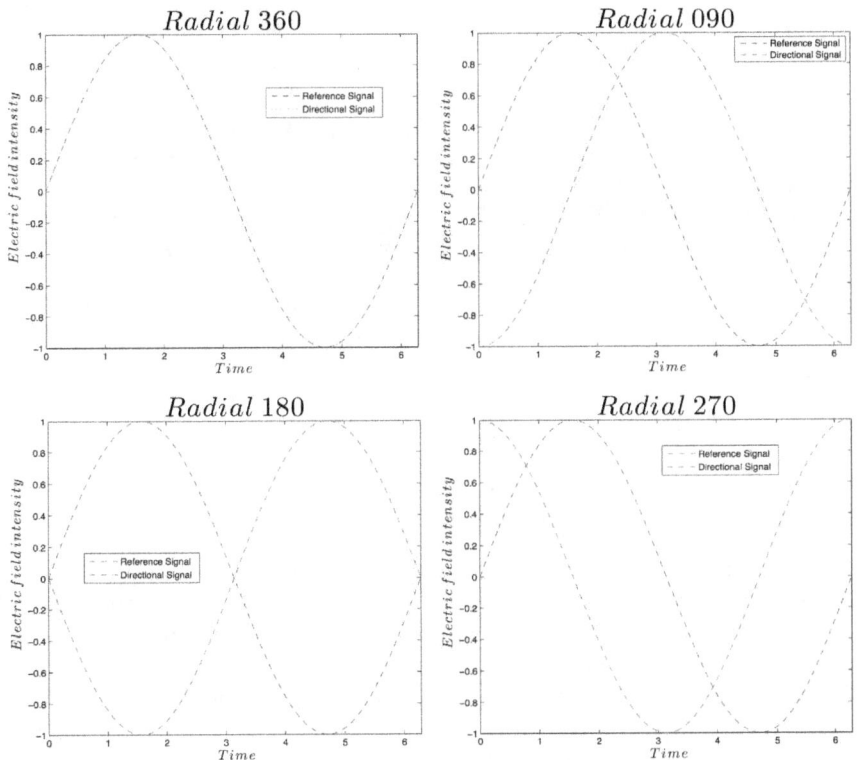

Figure 6.3: Omnidirectional and directional signals for different VOR radials

presentation on a VOR display does not depend on the heading of the aircraft. It depends only on the position of the aircraft with respect to the ground station and the selected radial.

VOR needles are centered when the selected radial is equal or opposite to the actual radial at which the aircraft is. A flag indicating TO or FROM is presented on the display to differentiate the outbound from the inbound track. The outbound magnetic track (FROM indication) from the VOR, when on a radial, will always be equal to the degrees of the radial at which the aircraft is. The inbound track will be the opposite direction (TO indication). Figure [6.5] shows different aircraft positions and associated VOR displays.

The VOR does only indicate the radial with respect to the ground station at which the aircraft is. It does not determine the distance to the station. Only when overflying the VOR, the TO/FROM flag will change. For a complete determination of position, a DME is normally associated to the VOR.

Figure 6.4: Conventional (left) and Doppler (right) VOR ground stations
© ZabMilenko & Yaoleilei / Wikimedia / CC BY 3.0 & CC BY-SA 2.0

3 Instrument Landing System

An ILS is an instrument navaid used for the development of a precision approach to a given runway threshold. The ILS signal has two components:

- A signal used for lateral navigation named *localizer* (LOC) which works similarly to a VOR, with higher accuracy requirements and lower range. The LOC ground antenna is placed at the end of the runway it serves. Surfaces of position are vertical semiplanes, similarly to the radials of a VOR, and define a nominal trajectory aligned with the centerline of the runway. The LOC signal is used by the pilot to know whether the runway centerline is to the left or to the right of his position.

- Another signal for vertical guidance named *glide slope* (G/S), which also works through phase comparison between one omnidirectional (reference) and multiple directional signals (different for each elevation angle from the threshold). In this case, the surfaces of position are defined from the runway threshold, perpendicular to the ones defined by the LOC signal. A nominal trajectory imposing a descent gradient of 3 degrees is typically defined. The G/S signal indicates whether the aircraft is flying above or below the nominal trajectory.

This navaid enables instrument approaches to a runway threshold even under poor visibility conditions. Figure [6.6] shows the ground equipment used to generate the two signals.

The instrument onboard displays both, the LOC and G/S relative position of the aircraft with respect to the nominal trajectory, helping the pilot guide the aircraft. Figure [6.8] shows an ILS onboard presentation.

4 Distance Measurement Equipment

All the previous navaids do not output a direct measurement of the aircraft´s distance to the ground transmitter. For this reason, a DME is normally attached to other ground navaids, providing full position awareness.

DME works (similarly to SSR systems) through an asking/answering process. Aircraft send electric pulses through DME transmitters that are received by ground DME equipments, which after 50 μs answer back other similar pulses that are captured by the aircraft´s DME receiver. Through the proper treatment of these signals, analyzing the total amount of time taken to get an answer after a question, the system onboard can determine the straight-line distance between the aircraft and the DME ground facility (see Figure [6.9]).

Each DME station has two assigned channels, one for reception and another one for transmission. These channels are defined every 1 MHz in the 962 - 1215 MHz frequency band.

Through the use of two different DMEs and an altimeter, one can try to determine the actual position (x_P, y_P, z_P) of the aircraft (see Figure [6.10]). A system of three equations may be set, obtaining a couple of solutions, one of which would not be valid.

$$\begin{cases} (x_P - x_1)^2 + (y_P - y_1)^2 + (z_P - z_1)^2 = (DME_1)^2 & First\,DME \\ (x_P - x_2)^2 + (y_P - y_2)^2 + (z_P - z_2)^2 = (DME_2)^2 & Second\,DME \\ z_P = h & Altimeter \end{cases}$$

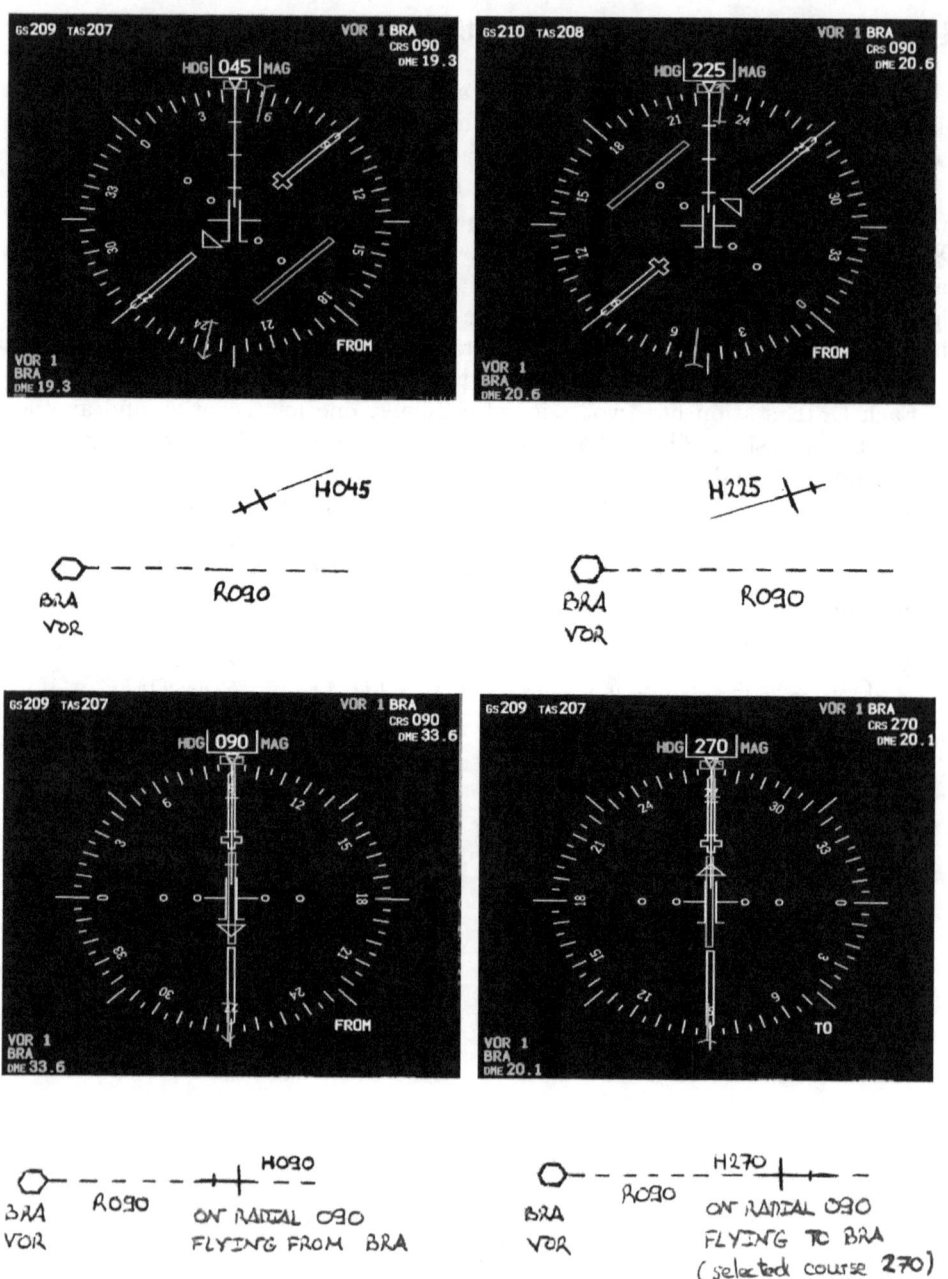

Figure 6.5: Position with respect to a VOR and instrument display

Figure 6.6: Localizer (left) and Glide Slope (right) ground stations © Herr-K / Wikimedia / CC BY-SA 3.0

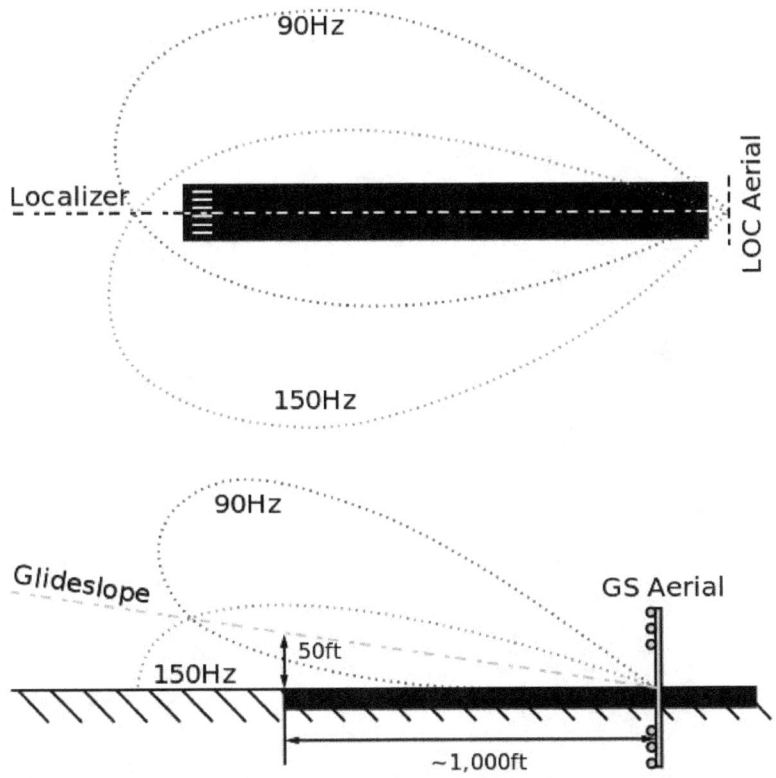

Figure 6.7: ILS signal detail and nominal trajectories © Fred the Oyster / Wikimedia / CC BY-SA 4.0

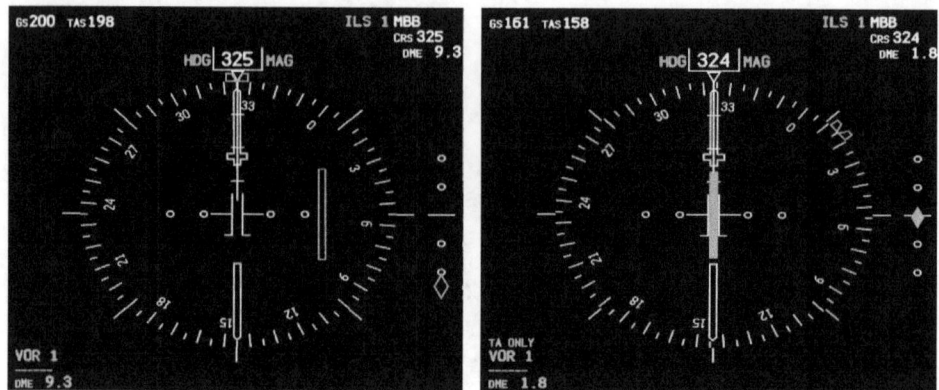

Figure 6.8: ILS instrument presentation - Too high and too left-deviated approach (left picture) and at the nominal trajectory (right picture)

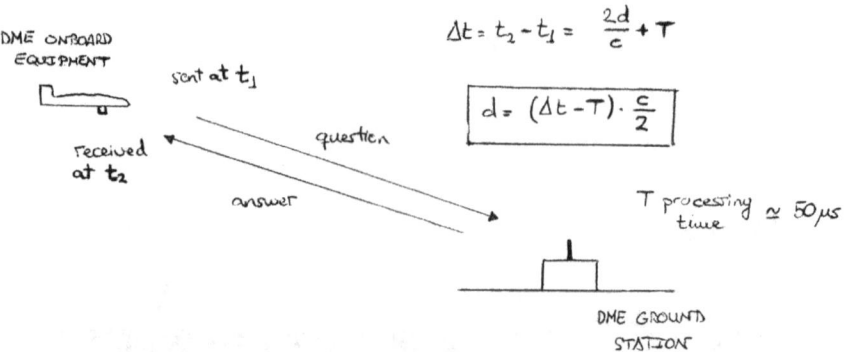

Figure 6.9: Telemetry principle for distance determination used by a DME

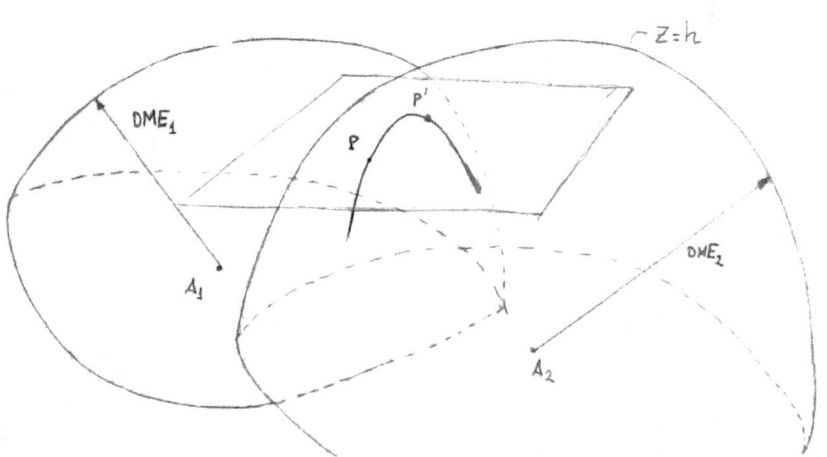

Figure 6.10: Position determination with two DME stations and an altimeter

Chapter 7. Space Navaids

In the previous chapter we have seen how ground stations can be used to send signals that support aircraft in the determination of their local position. However, in an environment where there is a need for global navigation, one would think of navaids that are available everywhere, and here is where satellites come into place. Satellite systems are used to support the determination of global position as space-based navaids.

The network of operating satellites that support the determination of position form the so-called GNSS (Global Navigation Satellite System). Different countries are involved in the development and exploitation of this network. The first, most complete and most reliable of all of them is the american GPS, to which most of this chapter is devoted.

By September 2014 there were 89 navigation satellites, a number which is expected to increase up to 150 by 2020. One may wonder why there are so many, whether some of them are redundant. We cannot forget that defense purposes rely in the origin of most satellite positioning systems. Countries want to have their own sovereign system, even though most of these systems have a dual use, civil and militar. For military reasons, it looks evident that a country will not rely on a satellite system being run by another country. As we will see through this chapter, from a civil perspective higher availability means higher accuracy, specially for situations where lines of sight can be blocked (buildings in cities) or multipathed (reflections). These redundancies make the system safer and more robust, since sometimes constellations have issues with one of their satellites. A constant double checking and the availability of multiple satellite sources also provides jam immunity.

The principle under which GNSS works is based on the determination of distances between satellites and users through the measurement of the time it takes for a signal to travel from the transmitting satellite to the receiving user. A common and very precise time reference frame is required. As we will see through this chapter, the emitted signals contain relevant information about the exact moment and position from where they are transmitted.

Figure [7.1] shows a single satellite clock and the most relevant information required for the determination of the range between the satellite and the user. With the information retrieved from one single satellite, a user would be in any point of a sphere centered on the satellite at the computed radius. An intersection of three spheres results in a point in space, so ideally we would require the information from three different satellites to solve for the position of the user. However, since the clock of the user is by definition inexpensive and not as accurate as the GPS Time Reference System, another unknown has to be added to the problem: the time difference between the satellite and the user clock. A total of 4 satellites is thus required to solve for the position and clock offset of the user. This is named "*estimanda*" and is defined as the position of the user and clock offset with respect to the GPS system time (x_u, y_u, z_u, b_u).

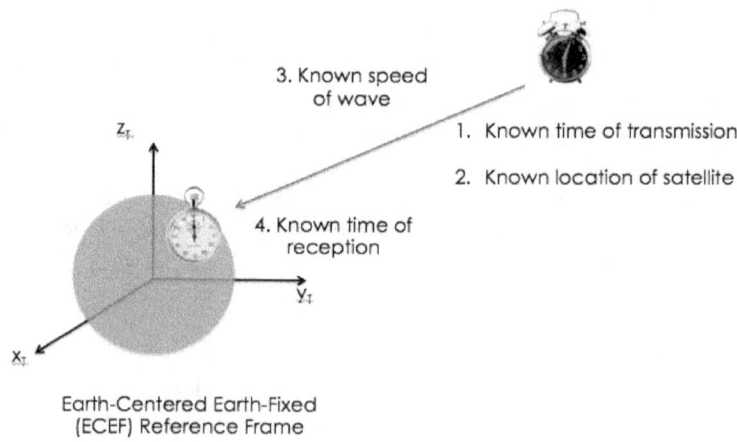

Figure 7.1: Basics of GPS

The existence of GPS has enabled accurate positioning and flexible aircraft navigation in regions where it was previously impossible, such as the approach to Juneau (Alaska) through a very narrow canyon. This channel was impossible to navigate before GPS. Now one can navigate within the canyon even if there is fog or low cloud ceiling.

Besides aviation, every smartphone in 2014 already had a GPS receiver, which was within a chip only 2 milimeters wide. By 2014 original GPS prices had already strongly fallen down, bringing GPS navigation to the public (car navigators, Google Maps, running & picture location applications, etc). While in the past navigation was driven by elites suchs as army officers, current development of technology is bringing navigation to the masses.

We will take a look within this chapter at Differential & Assisted GPS (DGPS & AGPS). These are techniques that enhance the precision in the determination of position.

We cannot forget that we have clocks in space moving very fast, so the different relativity theories have to be applied. According to general relativity, when gravity decreases time goes faster. Time on the orbit of a GPS satellite happens $45\mu s$ faster per day than on Earth. Also, according to special relativity, when travelling very fast (satellites move at around 3 kilometers per second with respect to the surface of the Earth), time goes $7\,\mu s$ per day slower. Gravity and speed variations happen along the orbit of the satellite because of its eccentricity (making it not circular). These relativistic offsets are programmed before the satellite launch, making the clocks slower (working at the expected time rate when airborne). Time variations because of eccentricity are corrected in the receiver code.

1   Satellites

Satellites are the primary infrastructure, since they are the ones that send the radio signals. Even though satellites travel at around $3\,\frac{km}{s}$ with respect to the user, they are so far away describing their orbits that one satellite may remain hours in the sky over the user, from the time it raises up over the horizon to the time it sets.

As we already stated in the introduction, four are the different enablers of GPS:

- Knowing the Time of Transmission (ToT) - very accurate atomic clocks are set onboard

- Location from which the signal is transmitted - the position of the satellite along time has to be computed with high precision, using orbital science. The data that describes the movement of the satellites is named *Ephemeris*

- Knowing the speed of the wave as it travels from the satellite to the user

- Measuring the Time of Arrival (ToA) - the arrival time is timestamped by the inexpensive clock of the receiver

Low Earth Orbit (LEO) satellites are used for surveillance or observation. The footprint of a LEO satellite is small. GPS could work on LEO but the constellation would need a lot more satellites (the visibility of each satellite is lower). Geostationary Earth Orbit (GEO) satellites are used when we want a satellite to be fixed to a given point on the surface of the Earth (it rotates at the same speed of the Earth). Communication satellites are typically GEO satellites. Also, satellites used for GNSS augmentation are GEO satellites, such as the WAAS & EGNOS constellations. GPS constellation of satellites operates in the Medium Earth Orbit (MEO, at around 20000 km of altitude), between LEO and GEO satellites. All GPS satellites are in a set of orbits named the GPS birdcage.

The US Air Force guarantees a minimum of 24 operative GPS satellites. However, since 2004 and at least until 2015, there have been at least 30 operative GPS satellites.

1.1 GNSS Orbits

As the reader may already know, it is thanks to the observations and calculations made initially by people like Brahe, Kepler or Newton that the science of orbital mechanics was developed. Basically, Brahe collected some data regarding the position of different celestial bodies, Kepler analyzed this data and Newton provided the model.

The determination of the position of the satellite (through the use of the tools provided by orbital mechanics) is essential since the user will be receiving signals from satellites and needs to know where these signals are coming from.

Since this is not an orbital mechanics text, we will just summary the main characteristics of the GPS orbits, assuming the reader is already familiar with the main parameters that define an orbit (particularly an ellipse).

The Fundamental Orbital Differential Equation for a two body problem (that is, assuming that the only two bodies affecting each other are the Earth and a single satellite) is $\ddot{\vec{r}} + \frac{\mu\vec{r}}{r^3} = 0$, where $\mu = M_e G$, which is solved by solutions of the type $\vec{r} = \frac{a(1-e^2)}{1+e\cos\nu}\vec{u}_r$. These solutions are ellipses where ν (named the *True Anomaly*) is the angle formed between the line that connects the Earth and the satellite and the perigee. The Earth is placed at one of the focal points of the ellipse. We describe the position of the satellite (see Figure [7.2]) with only six keplerian parameters[1]:

- Semimajor axis "a" of the ellipse that shapes the orbit

- Eccentricity "e" of the ellipse. The GPS orbits are very close to circular, however they are not perfect circles, they have eccentricity

- Angle between the Vernal Equinox[2] (inertial reference) and the ascending node Ω

- Location of the perigee through the "argument of perigee" ω

- Pitch the ellipse makes with respect to the equatorial plane, named *inclination* "i"

- True anomaly ν to place the satellite in the orbit

GPS satellites are organized into 6 different orbits. These 6 different orbits differ mainly in their respective Ω.

---

[1]Which are transmitted in the Navigation Message

[2]The Vernal Equinox is defined as the line that joins the center of the Sun and the center of the Earth during the spring and fall equinox. It points towards the Aries constellation.

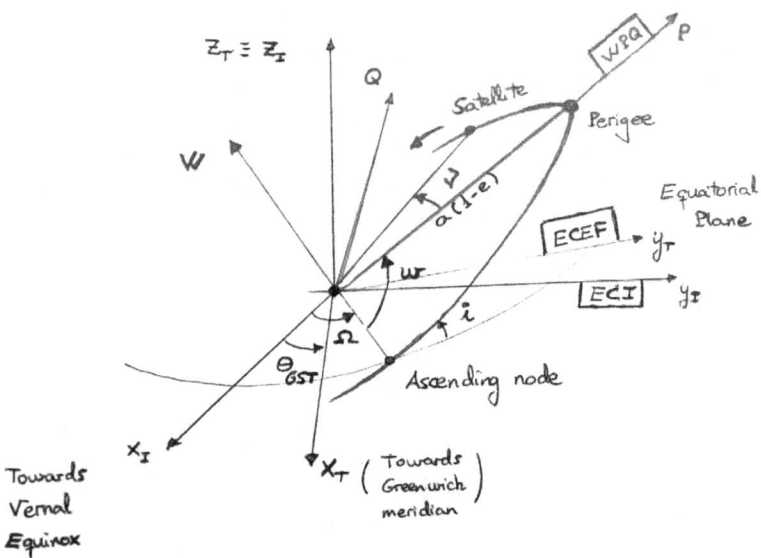

Figure 7.2: GNSS elliptical orbit and coordinate frames

1.2 Coordinate frames

For the description of the different movements, one may use one or another system of reference (to his convenience). We here analyze which are the most common (see Figure [7.2] again for further insight)[3].

The *Perifocal* (WPQ) reference system in which OP points towards the perigee of the orbit, OW is perpendicular to the orbit of the satellite and OQ makes a right-handed reference system.

The *Earth Centered Inertial* reference system (ECI) is an inertial frame in which OX points towards the Aries constellation, OXY sits on the Equator plane and OZ is the rotation axis of the Earth.

The *Earth Centered Earth Fixed* reference system (ECEF) is a reference frame attached to the Earth, pointing the OX axis towards the Greenwich meridian, sitting OXY on the Equator plane and OZ pointing towards the Geographic North. Most users are interested in this framework because they live attached to it.

The *East North Up* reference system (ENU) is used locally by a user placed on the surface of the Earth.

Most GNSS orbit calculations are transformations from one frame to another. Typically the process goes from Keplerian parameters → Perifocal reference system → ECI → ECEF

It is in the ECEF reference frame where we define the geographic coordinates λ

---

[3]All of them being right-handed reference frames

and φ. ECEF rotates attached to the Earth while ECI does not. The angle formed between ECI and ECEF is named θ_{GST} (GST stands for Greenwich Siderial Time) and is the angle formed between the meridian that contains the city of Greenwich and the Vernal Equinox. Be aware that we are measuring θ_{GST} with respect to the far stars, not with respect to the Sun. We do not talk about solar time (24 hours in a solar day), instead we talk about sidereal time (23 hours and 56 minutes in a sidereal day). One Solar Day = 1.0027379 Sidereal days.

2 Navigation messages

Navigation messages include information regarding who, when and from where was the radio signal broadcasted by the satellite. The information regarding position (*Ephemeris*) and time is sent by each satellite at a rate of 50 bits per second (which seems extremely slow when compared to WiFis that work at hundreds of millions of bits per second). Data speed is so low because coming from tens of thousands of kilometers away, the message is already quite weak by the time it reaches the user. Data compression is used to fit in all the required information into such a low rate of data.

The GPS Ground Control Segment uploads corrections to the *Ephemeris* and clock offsets of the satellites referred to a common GPS time, which is set by the Master Stations. The location of these ground stations is very well known and the GPS problem is reversed to solve for information about the position of the satellites. Messages are put together and uploaded to each satellite determining its position. The Ground Control Segment continuously tracks all GPS satellites, estimating clock offsets and positions for all of them. Infrequently the Ground Control Segment has some other critical functions such as command small maneuvers to maintain the orbit, command small clock corrections or command major relocations of the satellites because of any satellite failures.

The data in the message is contained in 25 frames, divided into five subframes each with an extension of 300 bits (6 seconds at 50 bps). A message thus contains 37500 bits and it takes up to 12.5 minutes to broadcast it. Subframe One contains information about the satellite such as its technical state and accuracy. Subframes Two and Three contain the *Ephemeris*. Subframe Four and Five contain the *almanacs* which are pieces of information regarding the position of other satellites in the constellation.

Satellites use signal carriers to send both:

- The Code (PRN) - which identifies the satellite

- The Data - position and clock offset of the satellite

2.1 Decoding the GPS navigation message

The *eccentric anomaly* (E) is defined relative to the center of the ellipse, with a virtual position of the satellite with the same abscise than the true one, in a circle of radius "a" (semi-major axis). See Figure [7.3] for a clearer definition of the *eccentric anomaly*.

The *mean anomaly* (M) is the angle at which the satellite would be if the angular speed were constant and equal to $n = \frac{2\pi}{T} = \sqrt{\frac{\mu}{a^3}}$. $M = M_0 + n(t - t_0)$. Only if the orbit were completely circular (no eccentricity), then we could say $\nu = M = E$.

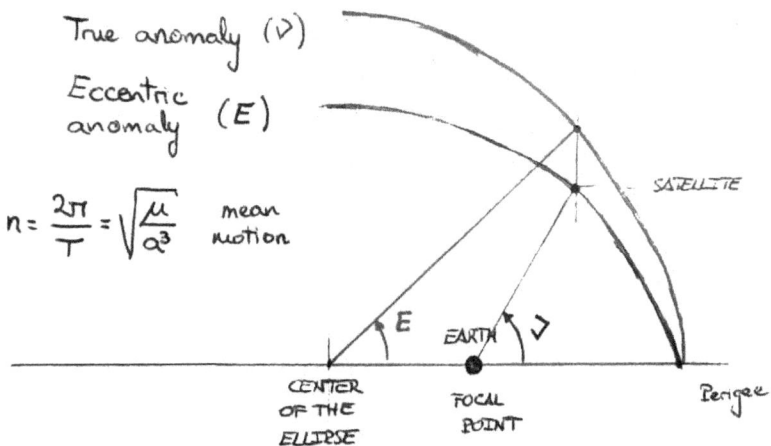

Figure 7.3: Eccentric Anomaly

From the GPS signal one actually gets the mean motion (M), we use it to get E since we know that $M = E - e\sin E$ and later get $\nu = \arctan\left(\frac{\sqrt{1-e^2}\sin E}{\cos E - e}\right)$.

As already said, the gravitational force to the Earth dominates when compared to other bodies such as the Moon, the Sun, etc. However, for a deeper understanding of the orbit of a satellite, these bodies have to be taken into account as well as the lumpiness of the gravitational Earth field. Also, effects of solar radiation and associated pressure have to be taken into account when analyzing the forces affecting the movement of the satellite.

In the navigation messages, some other parameters (besides the 6 that describe the position of the satellite) are included to take into account perturbations and their influence on the orbit.

3 Navigation signals

The navigation signal includes the carrier, the code and the data. The carrier is used for broadcast purposes and inside the message is coded (to determine from which satellite it comes from) and modulated. A satellite may send up to three pairs of signal + message, on three different bands (L1, L2 and L5).

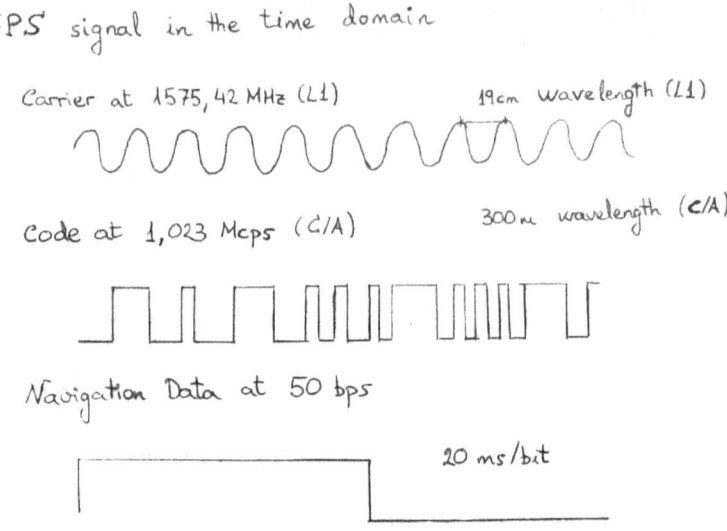

Figure 7.4: Example of a L1 GPS signal

The carrier signal is that part of the signal that carries the message from the radio transmitter to the radio receiver. It is characterized by its frequency or its wavelength. Most satellite signals work between 10^9 and 10^{10} Hz. This is because in that window the Earth's atmosphere is transparent. Below 10^9 Hz the ionosphere becomes more and more opaque to radio. Beyond 10^{10} Hz the wavelengths get shorter and they become more affected by the oxygen and water vapour in the atmosphere.

The PRN code is what identifies a satellite. This code includes binary chips, each of which is 300 meters wide in the L1 band. There are 1.023 Mega chips being clocked out per second. This is also named the Coarse / Acquisition code. There is a unique code for each GPS satellite. Among other things it can be used to identify to which satellite one is "listening" to. The receiver has a set of replicas, a book in which the codes of all the airborne satellites are stored, and through comparison (correlation), the receiver checks which is the satellite he's receiving messages from. These are pseudo-random codes that may look random but have a very carefully crafted structure, that supports quick correlation. Codes from one satellite are orthogonal to codes from other satellites, so one can do this autocorrelation one satellite at a time without having to worry too much about the signals coming from other satellites. The codes have been designed to have very sharp correlation peaks with themselves (autocorrelation) and very low correlation with any of the other satellite signals (cross-correlation).

GPS satellites use carriers in the L band, which goes from 1 GHz to 2 GHz. Some current (and all future) GPS satellites transmit in three different frequencies. Each of these frequencies includes the code, the data and the carrier. They are L1 (1575.42 MHz), L2 (1227.60 MHz) and L5 (1.176 GHz).

This frequency redundancy is good to mitigate other radio signals interference, ionospheric effects (as we will later see) and the fact that because the GPS signal is very weak (arriving with a strength of around $10^{-16} W$) we like having backup frequencies.

Different GNSS systems use different frequency bands which are stored in the spectrum in the so-called RNSS (Radio Navigation Satellite System) and ARNS (Aeronautical Radio Navigation System) bands. This will be more broadly explained in section (8).

3.1 The L band - Frequency domains

We have got the carrier, then the code modulated on the top of that, and finally the most slow, patient process which is the navigation data.

Atmospheric opacity is a function of the frequency of the signal that tries to go through. Between 100 MHz up to 20 GHz the opacity of the atmosphere drops. The atmosphere is transparent at these frequencies. The ionosphere may delay the signals within these frequencies, but not block them.

The L band was properly chosen taking int account all these effects. Lower frequency values would have larger ionospheric delay. Higher frequency choices would have more trouble with interference from human-made noise and would make the radio signals more sensitive to rain drops.

For rectangular pulses the sync function (that shapes the frequency domain) looks like $\frac{\sin x}{x}$. This shape in the frequency domain is moved to $\pm f$ where f is the frequency of the carrier.

4 Pseudoranging

It is the essential description of the measurements made by the equipment of the user. Internal replicas of the PRN code are moved back and forward until coincident with the received signal, it is then when correlation occurs and the Time of Arrival of the message (ToA) is stamped. Satellite stamps the transmission time (ToT) included in the message. Both ToT and ToA are stamped by different clocks. ToT are all controlled by the GPS System Time, but the ToA depends on the very inexpensive clock and will thus be biased (we will name it t_u when it is biased and ToA when it is unbiased). If there were no clock errors, we would say:

$$ToA - ToT = \frac{d}{c}$$

One satellite defines a sphere of possible locations. One can do the same thing with some other satellites. Three satellites would be enough for three dimensions to determine the point of intersection between three spheres. But due to the clock offset, we should know that if the user clock is fast, all ranges would be measured long and if the user clock is slow, all ranges would be measured short. The pseudorange differs from the true range basically because there is a clock offset.

$$t_u = ToA + b_u$$

where t_u is the time marked by the clock of the user and b_u is the clock offset. The GPS System does not have any control over the clock of the user. It is thus necessary to make a *state augmentation*, increasing the number of unknowns. We do this introducing a new unknown (b_u), a new equation and thus a new satellite. GPS works in a 4 dimensional space, we have to solve for $\{x_u, y_u, z_u, b_u\}$. Be aware that b_u does not depend on the satellite we are looking at. Every aircraft should thus be able to see at least 4 satellites. When solving, we also get b_u and thus we know the real GPS time. There are many applications only interested in retrieving the GPS time from the signal. However, the four satellites should not be very close to each other (the provided equation would not be independent). Normally 7 or 8 satellites in view are required.

$$t_u - ToT = \frac{d}{c} + b_u$$

Satellite number is a superscript. For the "j^{th}" satellite, the pseudorange equation goes as follows. Be aware that, just for quicker notation criteria, all terms must be unit-consistent (we do not write the speed of light everywhere).

$$\tau^{(j)} = \sqrt{\left(x_u - x^{(j)}\right)^2 + \left(y_u - y^{(j)}\right)^2 + \left(z_u - z^{(j)}\right)^2} + b_u + \epsilon_u^{(j)}$$

The *ephemeris* of the satellite must provide its position $\left(x^{(j)}, y^{(j)}, z^{(j)}\right)$. The error $\epsilon_u^{(j)}$ is typically of the order of one meter, as we will later model and analyze. Since the solution for these set of non-linear equations is complex, we will have to linearize them in order to get a system of equations that we can easily solve $\{x_u, y_u, z_u, b_u\}$.

4.1 Errors

Let us name $\nu_u^{(j)}$ as the error suffered by the user "u" in the measurement to the jth satellite. Errors have a normal distribution characterized by a null mean and a variation σ_τ^2. Figure [7.5] shows the different error sources:

- Satellite broadcasted and actual location differ. This error is generally within one meter. We characterize this error as a vector going from the actual to the transmitted location of the satellite

- The satellite makes a statement about its own time (error $B^{(k)}$). Difference between broadcasted and actual time is of the order of one nanosecond (one third of a meter)

- Reflections of the wave, that travel a route other than the one described by the Euclidean distance. These tend to be the greatest errors, leading normally to tenths of meters

- Ionosphere causes the wave to slow, relative to the speed of light. Error normally ranges from one to five or six meters when the signal comes from above. If the satellite were just over the horizon, the wave would spend a lot more time in the Troposphere, but not so much in the Ionosphere

- Troposphere also causes the wave to slow. Smaller error for satellites overhead, but larger when satellites over the horizon

- Natural noise

- Radio Frequency Interference (RFI) - man made noise

The largest error components when budgeting GPS errors for a stand alone receiver are normally the ionospheric delay and the multipath trajectory of the signal.

After we have done our very best to solve for the errors that we expect, we get the corrected pseudorange $\tau_C^{(k)}$ which is:

$$\tau_C^{(k)} = d_u^{(k)} + b_u - \delta B^{(k)} + \delta I_u^{(k)} + \delta T_u^{(k)} + \nu_u^{(k)}$$

Being the *user state* or *estimanda* $\overrightarrow{x_u} = (x_u, y_u, z_u, b_u)$ and the *state of the satellite* $\overrightarrow{x^{(k)}} = \left(x^{(k)}, y^{(k)}, z^{(k)}, B^{(k)}\right)$

Being $d_u^{(k)}$ the true range from the satellite to the user:

$$d_u^{(k)} = \sqrt{\left(x_u - x^{(k)}\right)^2 + \left(y_u - y^{(k)}\right)^2 + \left(z_u - z^{(k)}\right)^2}$$

4.2   Linearization

In order to solve the system of equations, we will first linearize them. For a one-dimension only problem, the reasoning goes as follows:

Satellite (k) is at $\overrightarrow{x^{(k)}} = \left(x^{(k)}, y^{(k)}, z^{(k)}\right)$
The Assumed Location is $\overrightarrow{x_{u,0}} = (x_{u,0}, y_{u,0}, z_{u,0})$
The Pseudorange Location is $\overrightarrow{x_u} = (x_u, y_u, z_u) = \overrightarrow{x_{u,0}} + \delta \overrightarrow{x_u}$

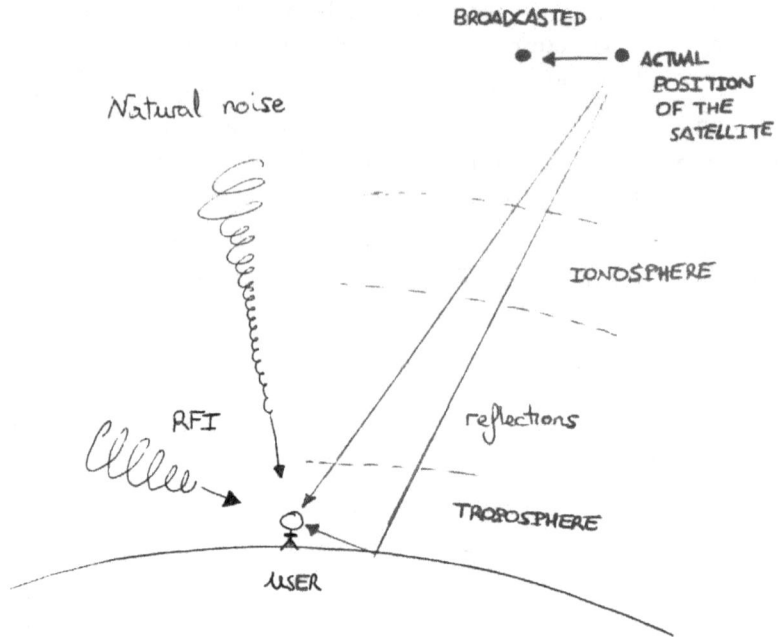

Figure 7.5: Error sources

We will assume a location that is hopefully near the true location and try to solve for the error in the assumption. If the first assumed location were very wrong, we would go iterating. GPS normally works in 3 or 4 iterations for the assumed location.

We will direct our energy at finding out a good estimate for $\delta \vec{x_u}$. First we compute the Euclidean distance between the broadcasted position of the satellite and the assumed location. We name this *theorange* $d_A^{(k,B)}$. The measurement is $\tau_u^{(k)} \approx d_u^{(k)} + b_u$. We substract these two parameters naming the *residual* or *innovation* $\delta\tau_u^{(k)} = \tau_u^{(k)} - d_A^{(k,B)}$. In a one-dimension problem, this distance is related to $\delta\vec{x_u}$ if we knew the elevation angle. We consider that the satellite is so far away that the lines that join the satellite with both the assumed and the pseudorange location are parallel to each other. See Figure [7.6].

$$\tau_u^{(k)} - d_A^{(k,B)} = -\delta x_u \cdot \cos\left(el_u^{(k)}\right) + b_u$$

Here we use an East - North - Up reference frame (ENU). The *estimanda* appears now in a linear form (δx_u and b_u). So far we have one equation and two unknowns. Just by adding up a new satellite, we would be able to solve for the two unknowns.

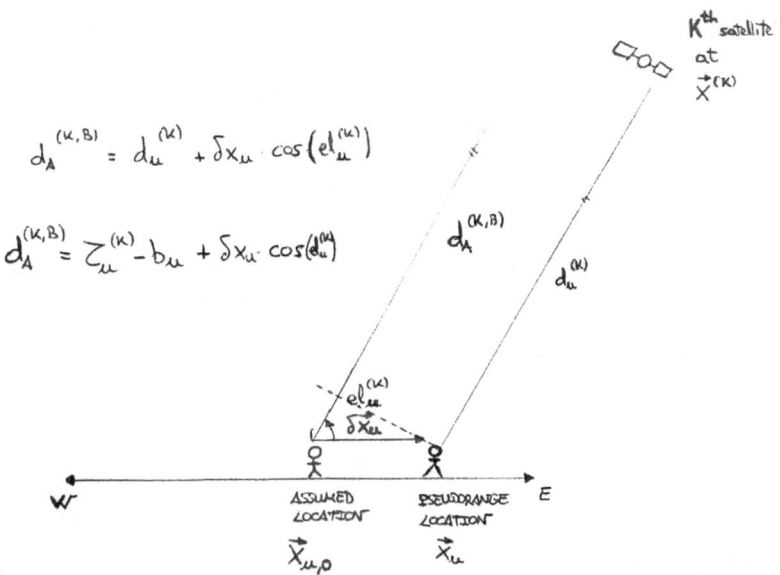

$$d_A^{(\kappa,B)} = d_u^{(\kappa)} + \delta x_u \cos\left(el_u^{(\kappa)}\right)$$

$$d_A^{(\kappa,B)} = \zeta_u^{(\kappa)} - b_u + \delta x_u \cos\left(el_u^{(\kappa)}\right)$$

Figure 7.6: Linearization for the one dimensional case

This can be extrapolated to a four dimensional problem with multiple satellite views. Defining the elevation angle as the angle made by the line of sight with the satellite with respect to the horizon and the Azimut angle as the one formed by the line of sight relative to the North.

$$
\begin{bmatrix} \delta\tau_u^{(1)} \\ \delta\tau_u^{(2)} \\ \delta\tau_u^{(3)} \\ \dots \\ \delta\tau_u^{(k)} \end{bmatrix} = \begin{bmatrix} -\cos\left(el_u^{(1)}\right)\sin\left(az_u^{(1)}\right) & -\cos\left(el_u^{(1)}\right)\cos\left(az_u^{(1)}\right) & -\sin\left(el_u^{(1)}\right) & 1 \\ -\cos\left(el_u^{(2)}\right)\sin\left(az_u^{(2)}\right) & -\cos\left(el_u^{(2)}\right)\cos\left(az_u^{(2)}\right) & -\sin\left(el_u^{(2)}\right) & 1 \\ -\cos\left(el_u^{(3)}\right)\sin\left(az_u^{(3)}\right) & -\cos\left(el_u^{(3)}\right)\cos\left(az_u^{(3)}\right) & -\sin\left(el_u^{(3)}\right) & 1 \\ & \dots & \dots & \dots \\ -\cos\left(el_u^{(k)}\right)\sin\left(az_u^{(k)}\right) & -\cos\left(el_u^{(k)}\right)\cos\left(az_u^{(k)}\right) & -\sin\left(el_u^{(k)}\right) & 1 \end{bmatrix} \begin{bmatrix} \delta E_u \\ \delta N_u \\ \delta U_u \\ \delta b_u \end{bmatrix}
$$

$$
\begin{bmatrix} \delta\tau_u^{(1)} \\ \delta\tau_u^{(2)} \\ \delta\tau_u^{(3)} \\ \dots \\ \delta\tau_u^{(k)} \end{bmatrix} = \begin{bmatrix} \tilde{G}^{(1)} \\ \tilde{G}^{(2)} \\ \tilde{G}^{(3)} \\ \dots \\ \tilde{G}^{(k)} \end{bmatrix} \begin{bmatrix} \delta E_u \\ \delta N_u \\ \delta U_u \\ \delta b_u \end{bmatrix}
$$

This is named the *geometry matrix*. Through this matrix we get the *residual* or *innovation* $(\overrightarrow{\delta\tau_u})$. We also get a notion of what the pseudorange should be if we moved the estimated position of the user in each of the East, North or Up directions. One may wonder, how coherent is our guess with respect to each of the satellites?

Diversity in the elevation and azimut angles is required to make a diverse system of equations.

If we used an ECEF reference frame, the geometry matrix would be different, and the system would look like:

$$\delta\tau^{(k)} = -\delta\overrightarrow{x_u} \cdot \underline{1}_u^{(k)} + \delta b_u + \tilde{\nu}_u^{(k)}$$

If we take 4 satellites and we assume that the noise is negligible, the geometry matrix will be a squared 4x4 matrix. Solving the linearized system of equations is as simple as inverting this matrix, multiplying later by the difference between the measurements and the estimations and get the *estimanda*.

For K = 4 (exactly specified case) we thus have:

$$\delta\underset{\rightarrow}{\tilde{x}_u} = G^{-1}\left(\delta\underset{\rightarrow}{\tau} - \underset{\rightarrow}{\tilde{\nu}_u}\right) = G^{-1}\delta\underset{\rightarrow}{\tau} - G^{-1}\underset{\rightarrow}{\tilde{\nu}_u}$$

For K > 4 (overspecified case). We have more equations than unknowns because we view more than 4 satellites. We get the minimum square error solution as:

$$\delta\underset{\rightarrow}{\tilde{x}_u} = \underbrace{(G^T G)^{-1}G}_{pseudoinverse}\left(\delta\underset{\rightarrow}{\tau} - \underset{\rightarrow}{\tilde{\nu}_u}\right)$$

A common technique is to weigh the different satellites depending on their position when the system is overspecified, removing those whose error is larger and more likely to bias the *estimanda*.

4.3   Dilution of precision

In this subsection we want to answer the question: how well does GPS work? and for this we will use the same set of equations. We try to connect how good the measurements are with how good the estimation of our position is. We should end up with the standard deviation (σ) of the error in the North - East - Up reference frame.

The geometry matrix determines how strongly the pseudorange measurement errors amplify into erros in X, Y, Z and T. We care about the variances[4] $\sigma^2_{x_u-\hat{x}_u}$, $\sigma^2_{y_u-\hat{y}_u}$, $\sigma^2_{z_u-\hat{z}_u}$ and $\sigma^2_{b_u-\hat{b}_u}$. These variables are normalized by the variance of the pseudorange[5] (σ^2_τ) and appear along the diagonal of matrix $(G^T G)^{-1}$ when we work in an Earth Centered Earth Fixed reference frame.

---

[4]Where $\vec{x}_u$ is the true state vector of the user and $\hat{\vec{x}}_u$ is the estimation of the state vector provided by the GPS receiver.

[5]We assume that all pseudoranges have the same variance associated with their error. We do not consider the fact that some satellites might have smaller variance than others.

$$
\left(G^T G\right)^{-1} =
\begin{bmatrix}
XDOP = \frac{\sigma^2_{x_u - \hat{x}_u}}{\sigma^2_{\tau}} & \ldots & \ldots & \ldots \\
\ldots & YDOP = \frac{\sigma^2_{y_u - \hat{y}_u}}{\sigma^2_{\tau}} & \ldots & \ldots \\
\ldots & \ldots & ZDOP = \frac{\sigma^2_{z_u - \hat{z}_u}}{\sigma^2_{\tau}} & \ldots \\
\ldots & \ldots & \ldots & TDOP = \frac{\sigma^2_{b_u - \hat{b}_u}}{\sigma^2_{\tau}}
\end{bmatrix}
$$

If instead we were working in an East - North - Up (ENU) reference frame, the geometry matrix we would use would be different and the Dilution of Precision comes determined by:

$$
\left(\tilde{G}^T \tilde{G}\right)^{-1} =
\begin{bmatrix}
EDOP = \frac{\sigma^2_{E_u - \hat{E}_u}}{\sigma^2_{\tau}} & \ldots & \ldots & \ldots \\
\ldots & NDOP = \frac{\sigma^2_{N_u - \hat{N}_u}}{\sigma^2_{\tau}} & \ldots & \ldots \\
\ldots & \ldots & UDOP = \frac{\sigma^2_{U_u - \hat{U}_u}}{\sigma^2_{\tau}} & \ldots \\
\ldots & \ldots & \ldots & TDOP = \frac{\sigma^2_{b_u - \hat{b}_u}}{\sigma^2_{\tau}}
\end{bmatrix}
$$

The dilution of the precision comes thus determined by the errors in the pseudorange and by the position in the sky of the satellite we are looking at. Not all satellite geometries are equal. Depending on their position, we will dillute more or less the precision of the estimated position of the user as shown on Figure [7.7].

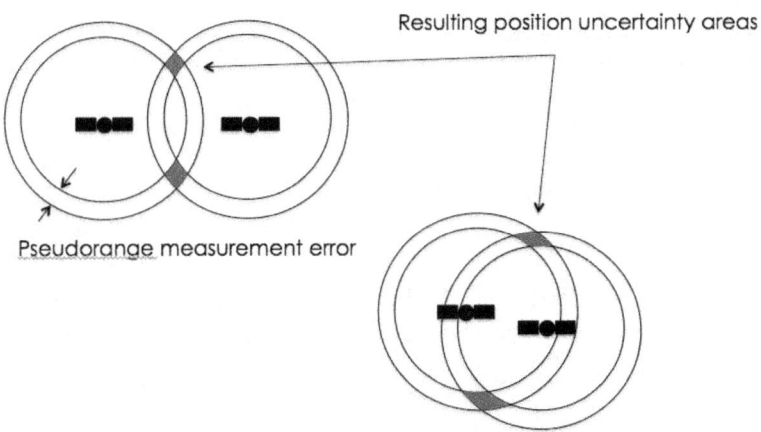

Figure 7.7: Dilution of precision as a function of geometry

This concept was used to design the constellation, taking into account that geometry matters. Different orbital planes and different satellites per orbital plane (computing their relative positions) were simulated to ensure minimum dilution of precision on any point on Earth at any time of the day. DOP is also used to select which satellites are the best to look at.

From an aviation perspective, we must take into account that when an aircraft is banking, it changes its elevation and azimut angles and thus the whole DOP matrix. This analysis must be performed in real time as the user equipment is being used.

5 Performance

In the determination of position, vertical errors are always larger than horizontal errors as we have already seen because of the structure of the DOP matrix. A normal stand-alone receiver has a horizontal precision of around 5 meters, that can be narrowed down to one centimeter with differential GPS. Based on the Doppler frequency shift, one can also estimate speeds with GPS. The frequency of the carrier shifts up when the range between the transmitter and the receiver is closing, and shifts down when the range is opening. Errors in GPS speed determination are of the order of $0.2 \frac{m}{s}$.

5.1 Differential GPS

A reference receiver whose position is very well known can receive GPS signals and estimate very well the bias of the signals. It can later broadcast this bias to all the users in the vicinity who, more or less, would be receiving similar signals. This way we know better which is the actual bias of the GPS signal and reduce the value of σ_τ. We can go from an accuracy of around five to ten meters with normal GPS to one meter accuracy using differential GPS. Figure [7.8] shows how DGPS works. We will solve where $\vec{x}_u$ is relative to $\vec{x}_R$. Note that $\vec{x}_u = \vec{x}_R + \Delta\vec{x}_{u,R}$.

We may recall that for a *stand alone* GPS receiver, we linearized using an initial guess of the position of the user, getting the *residual* by substraction. That is, initially we knew that the pseudorange for a kth satellite (as measured) is:

$$\tau_u^{(k)} = \|\vec{x}^{(k)} - \vec{x}_u\| + I + T + b_u - B^{(k)} + \nu$$

The theorange, computed from the broadcasted position of the satellite to the initial guess of position of the user is:

$$\tau_0^{(k)} = \|\vec{x}^{(k,B)} - \vec{x}_{u,0}\| + \hat{I} + \hat{T} + b_{u,0} - B^{(k,B)}$$

where $\hat{I}$ and $\hat{T}$ are estimations of the ionospheric and tropospheric errors. By substraction we got the *residual*, which is:

$$\delta\tau^{(k)} = -\Delta\vec{x}_{u,0} \cdot \underline{1}_u^{(k)} + \delta\vec{x}^{(k)} \cdot \underline{1}_u^{(k)} + \delta b_u - \delta B^{(k)} + \delta I + \delta T + \nu$$

Differential GPS has a very similar reasoning behind. The pseudorange equation (measurement at the receiver of the user) is identical:

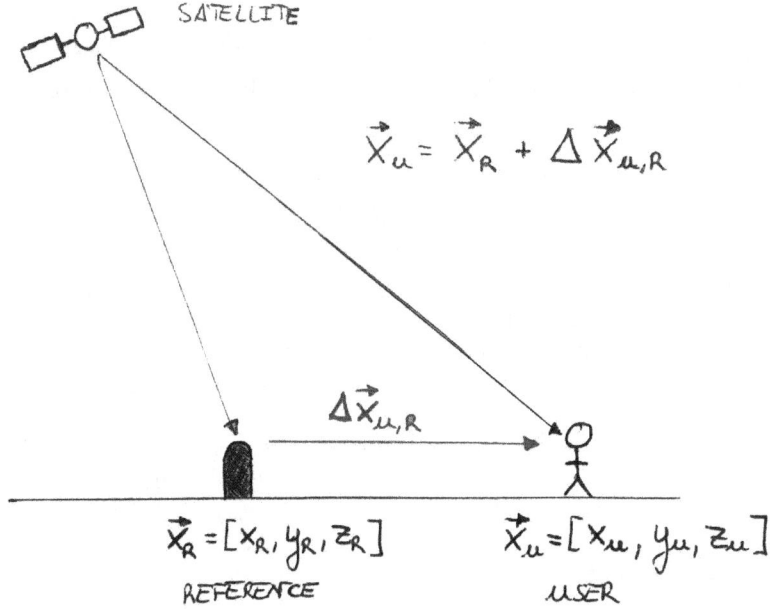

Figure 7.8: Differential GPS fundamentals

$$\tau_u^{(k)} = \|\vec{x}^{(k)} - \vec{x}_u\| + I_u + T_u + b_u - B^{(k)} + \nu_u$$

And now, instead of computing a theorange, we will replace it by a second pseudorange measurement taken at the reference station:

$$\tau_R^{(k)} = \|\vec{x}^{(k)} - \vec{x}_R\| + I_R + T_R + b_R - B^{(k)} + \nu_R$$

Be aware that now $\vec{x}_R$ and b_R are known since we know the state vector of the reference receiver to perfection. Substracting these two previous equations, we get the differential range equation that goes as follows:

$$\tau_u^{(k)} - \tau_R^{(k)} = \|\vec{x}^{(k)} - \vec{x}_u\| - \|\vec{x}^{(k)} - \vec{x}_R\| + \Delta I + \Delta T + b_u - b_R + \nu_u - \nu_R$$

ΔI and ΔT are the difference in the ionosphere and troposphere as experienced at the user and reference receivers. There is no modelling involved. We will assume that $\vec{x}^{(k)} - \vec{x}_u$ and $\vec{x}^{(k)} - \vec{x}_R$ are nearly parallel to each other. The mathematical reasoning is very similar to the one already described in section 4.2 regarding the trigonometry relationships and construction of geometry matrixes. Because of that, we can derive that $\|\vec{x}^{(k)} - \vec{x}_u\| - \|\vec{x}^{(k)} - \vec{x}_R\| = -\Delta\vec{x}_{u,R} \cdot \underline{1}_u^{(k)}$ and replacing it in the differential range equation we get:

$$\tau_u^{(k)} - \tau_R^{(k)} = -\Delta\vec{x}_{u,R} \cdot \underline{1}_u^{(k)} + \Delta I + \Delta T + \Delta b_{u,R} + \Delta\nu_{u,R}$$

For a multiple number of satellites, we can write the following set of equations:

$$\begin{bmatrix} \tau_u^{(1)} - \tau_R^{(1)} \\ ... \\ \tau_u^{(k)} - \tau_R^{(k)} \end{bmatrix} = G \begin{bmatrix} \Delta x_{u,R} \\ \Delta y_{u,R} \\ \Delta z_{u,R} \\ \Delta b_{u,R} \end{bmatrix} + \Delta I + \Delta T + \Delta\nu_{u,R}$$

Our earnest hope is that the reference and user receivers are close enough to each other so that ΔI and ΔT are small. We expect that the ray going from the satellite to the user goes through the same ionosphere and troposphere than the ray going from the satellite to the reference station. Thus, we can assume $\Delta I = \Delta T \approx 0$.

However, the differences in the random noises are not so easily cancelled since, for instance, reflection effects (multipaths) might not be similar in both, the user and the reference. We will thus have to stick with $\Delta\nu_{u,R}$ and expect to do something in the reference and user receivers to keep these random noises small. We use DGPS only to alienate the errors that are common to both, reference and user.

5.2 Errors in DGPS

We will first name those errors which are totally or mostly cancelled:

- Errors in the actual vs broadcasted clock time of the satellite are totally cancelled

- *Ephemeris* errors are reduced by DGPS but not totally cancelled. The projection of the $\Delta\vec{x}$ between the true and broadcasted location of the satellite over the rays that join the satellite and the user and reference station may not be exactly equal

- The closer the user and the reference station, the more similar the portions of ionosphere and troposphere the rays come through, experiencing the same delaying effects and thus almost cancelling the distortion in the determination of position

Other effects which are not mostly cancelled are:

- Multipath induced error - reflections caused by the immediate environment of the receiver

- Natural and man-made noise may tend to affect the reference station and the user very differently

5.3   Ionospheric effects on DGPS

During solar storms, big gradients in the effect of the ionosphere can be observed. If in a calm day with DGPS we may get values of ΔI of the order of 0.3 meters (for user and reference station separated about 200 km), during a storm day, for the same user and reference locations, we may get errors in DGPS due to ionospheric effects of up to 18 meters.

Storms in the ionosphere can lead to errors of up to tenths of meters in DGPS. In November 2003 a big solar storm happened and it became evident that new solutions were required for these events. All data so far had been retrieved from the L1 radio signal. The community looked for solutions to try to prevent this degradation of performance in the event of solar storms. The proposed solution was the use of dual frequency for GPS & GNSS in general.

5.4   Use of dual frequency

GPS frequency allocation as we already depicted ranges from 1.1 to 1.6 GHz. Out of the three billion GPS users worldwide, 99.9% of them do only use the L1 frequency band[6].

Different pseudorange measurements are taken at different frequencies to totally remove the effect of the ionosphere. Tropospheric delay does not change as a function of the signal frequency, but ionospheric delay does. This ionospheric delay is a known function of the frequency of the signal and models exist.

Thus, GPS moved towards a system in which satellites broadcast signals in three different frequencies (L1, L2, L5). Similarly have Glonass and Galileo done so far. These triple frequency satellites are currently being launched, so not all of the operative have this capability.

With these three frequencies one can measure the exact condition of the ionosphere at all instants through the comparison of the signals received from the satellite.

We can thus get an ionosphere-free signal at the user, without any assistance provided from the reference station. The performance of the stand alone user receiver improves dramatically because one of the largest error contributors (the ionosphere) can be cancelled out through the use of multiple frequency signal inputs.

---

[6]Be aware that most smartphones make use of the L1 frequency signal only

6 Power of the signal, Receiver design and Signal acquisition

6.1 Power of the signal

Each satellite transmits the GPS signal with a power of 27 W (about the same as a small light bulb, up in the space). This power spreads out in a sphere. The amount of power density in the sphere decreases as the sphere gets larger radius. Then we can find a power density function which is $p(r) = \frac{P_T}{4\pi r^2}$. For larger values of r the power density is so low that we instead talk about decibels (dBs).

Just as a quick summary we will remember that a "bel" is a ratio of powers. Power ratio in bels is defined as $\log_{10}(ratio\,of\,powers)$. Power ratio in decibels equals ten times the power ratio in bels. $dB = 10\log_{10}(ratio\,of\,powers) = 10\log_{10}\frac{P}{P_R}$. If $P_R = 1W$ we then talk about dBW. If $P_R = 1\,mW$ we then talk about dBm.

6.2 Antenna gain and reciprocity

The GPS satellite does not radiate power equally in all directions, it is more like a spotlight. It focuses all efforts in certain directions. The *antenna gain* is the ratio of the total sphere area to the spot (illuminated) area. For a given α (semiangle with which the satellite looks at the Earth), we can demonstrate that the gain of transmission $G_T(\alpha) = \frac{2}{1-\cos\alpha}$. For GPS orbits, if the satellite illuminated the Earth exactly and nothing more than that, then α would be 13.9º. However, a GPS satellite does not exactly illuminate the Earth, it has a wider view, the actual $\alpha = 21.3º$. This means that the signal illuminates the Earth and a little bit more.

GPS does not emit signals with the same power in all directions within the spotlight. GPS satellites send signals with a power such that all users receive similar strength values (further ranges require more powerful emitted signals).

Taking into account that a GPS satellite is designed with a $P_T = 27\,W$, an antenna gain $G_T = \{10\,to\,17\}$, then the effective power ranges from $270W\,to\,460\,W$ at the point of emission. There are two gains, G_T for the transmitter but also G_A for the receiver.

Since the $Received\,Power = \frac{P_T\,G_A\,G_T}{4\pi R^2}$, for GPS gives values of around $1.41\,10^{-16}W$ or $-128.5\,dBm$ on the surface of the Earth. No matter where the user is on Earth, the GPS system will try to provide him with the same signal power. This strength will also be a function of the receiving antenna.

The minimum guaranteed signal strength in an outdoor environment is $-130\,dBm$ for GPS.

7   Assisted GPS

Since the data coming from the satellites has very low rates (of 50 bps), a system has been designed to get relevant information (such as satellite positions) from different communication datalinks (that, instead, work typically at 1 Mbps). When looking for a satellite, the user has to search in the frequency domain for the signal (which will be around the L1, L2 or L5 frequencies, plus the associated Doppler effect because of the relative motion between the satellite and the user). If the position of the satellite is already provided through the use of a communication datalink, the range of possible frequencies in which the user expects the signal to be is narrowed. The user reduces thus the search space in advance, for both (f, τ) variables.

The initial guess of the user about his position (in the vicinity of the antenna from where he is getting the communication information) and the fact of knowing in advance the *ephemeris* of the satellite makes the process of (f, τ) search a lot shorter. That is why AGPS is sometimes called "Instant GPS".

AGPS does also support higher sensitivity and thus, indoor GPS.

If the user is going to work with GPS indoors, he will need to apply *non-coherent integration* techniques (which allows for longer integration times). The power of the received signal is 1000 times weaker when indoors, but integrating for longer we can get the signal over the noise, with a Signal to Noise Ratio of around 17 dB.

The proliferation of cheaper receivers together with AGPS has enabled that all smartphones currently built include a GPS receiver. There are thousands of applications that make use of GPS signals even for indoor navigation.

8   GNSS

GNSS stands for Global Navigation Satellite System and is an inclusive term of the different projects developed by certain world regions or countries, which are:

- USA -> GPS

- Russia -> Glonass

- Japan -> QZSS

- China -> Beidou

- European Union -> Galileo

- India -> IRNSS

Each of them is characterized by the designing and operating region or country, the required orbits and the signal and message frequency and encoding.

8.1 GPS

GPS was the first GNSS system and designers chose an orbit so that the period of each satellite to be half a siderial day. Each satellite flies the same orbit twice each sidereal day (faster than a GEO satellite). With this design, everything repeats every sidereal day. This is something special for GPS and different for all the other systems. Satellites are placed on 6 different planes (resting an average 5 satellites per plane). These orbits have an inclination of 55 degrees. The projection of the position of the satellite on the ground, as the Earth rotates, draws a line similar to a tennis ball.

Initially Block IIA/IIR satellites were used transmitting civilian signals only in L1. Current smartphones and most civilian receivers operate in this frequency only. Later, Block IIR-M were launched sending civilian signals also in L2 (for dual frequency receivers). Finally, Block IIF satellites were also sent including L5 signals. All of these use BPSK encoding. Current GPS satellites last over 10 years.

In the future GPS-3 (Block III of satellites) will be launched in which not only all three L1, L2 and L5 frequencies will be used, but also the modulation will be performed differently using BPSK (PRN code multiplying the carrier wave so the phase changes every now and then) + BOC (Binary Offset Carrier - in which one PRN code is placed over another one). BOC gives shape to a more complicated frequency spectrum (with two main lobes) instead of the classical sync function. One can have the PRN code and the BOC code without interferring with each other.

8.2 GLONASS

It is the russian system. It was the first fully operational system after GPS. Glonass is at a slightly lower orbit than GPS (19100km versus 20200 km). The orbital period is $\frac{8}{17}$ sidereal days. This means that the movement is repeated every 8 siderial days. There are 24 satellites in 3 planes (8 satellites per plane). Inclination of the orbits is 66 degrees to the equator. Glonass has only one PRN code (not one for each satellite as happened in GPS). We recognize the source of the signal because each satellite transmits in a different frequency. This is named FDMA (Frequency Division for Multiple Access). Future Glonass-K will support FDMA and CDMA (it is going to be more like GPS, using different PRN codes). Each of the 24 satellites sends signals centered on two main frequencies (named L1, L2). A third Glonass L3 lower frequency will work for in the future Glonass-K.

8.3 QZSS

QZSS stands for Quasi-Zenith Satellite System. It is the japanese system and was launched by Jaxa, the japanese space agency. The orbits are in the GEO distance, but some of them are inclined. The orbit is not symetric, with high eccentricity, so the satellite spends more time over Japan than over the rest of the surface of the Earth. Orbital planes have an inclination of 41 degrees. The period of the satellites is one sidereal day. Altitude of the orbit is 35786 km. There is currently just one satellite (named *Michibiki*), but they expect to finally include 4 IGSO and 3 GEO satellites. The signal is similar to the one used by GPS, making it easy to modify a GPS receiver to be compatible with QZSS. QZSS is already providing the L1C signal (with two lobes in the frequency spectrum). The japanese are also making experiments in LEX (frequency between L1 and L2, that will be named L6 in the future), which will provide centimeter accuracy within the region of Japan.

8.4 BeiDou

It is the formerly called Compass chinese system. BeiDou has 5 GEO, 5 IGSO and 4 MEO satellites. The MEO satellites operate similarly to GPS (providing global coverage), they operate at an altitude of 21500 km (slightly higher than GPS satellites) and with an orbit period of $\frac{7}{13}$ (the constellation repeats its position every 7 sidereal days). The orbits of the 4 MEO satellites have an inclination of 55 degrees.. Initial plans are to go up to 27 satellites by 2020. It is characterized by having a very sparce coverage away from Asia and a very good coverage over Asia. Signals are CDMA transmitted (similarly to GPS) in three different frequencies: B1 (1562 MHz), B2 (1207 MHz) and B3 (military signal). B1 is the only frequency used by smartphones. In the future, they expect to move into the so-called BeiDou-3, with further global coverage to be provided by the new satellites.

8.5 Galileo

It is the navigation satellite system operated by the European Union. Galileo stays at MEO (23200 km of altitude). It is planned to have 30 operational satellites (currently only 3 deployed) distributed along 3 different orbital planes (10 satellites per plane) with an inclination of 56 degrees. The period of these orbits is $\frac{10}{17}$ sidereal days, meaning the constellation repeats its position every 10 days. In september 2014 the first fully operational satellites were being launched and were finally placed at lower orbits than expected. Galileo has been designed to be interoperable with GPS. Signals are sent in three different frequency bands, out of which two (E1 & E5a) exactly match their GPS equivalents (L1 & L5). The BOC encoding that is going to be on GPS-3 is already on the Galileo signals. Initially there were two test satellites (GIOVE A & B) which are no longer operational. There were also 4 In Orbit Validation satellites (IOV) which were test satellites out of which three still remain operational. There are also two Fully Operational Vehicles (FOV) which are the failed-launch satellites (remained at a lower orbit). Similarly to the japanese, satellites get the names of those children that win drawing competitions. The two FOV satellites are going to be placed into a better orbit to reduce the Doppler deviation frequencies currently received on Earth (which double GPS values).

8.6   IRNSS

Last but not least we have the Indian Regional Navigation Satellite System. It exists at the GEO and GSO altitudes. The plan is to have 3 GEOs and 4 GSOs (intended to produce symmetric 8 shapes on the surface of the Earth over India). By September 2014 they only had two GSOs (IRNSS1A launched in July 2013 and IRNSS1B launched in April 2014). Unlike all the other GNSS constellations, it does not emit any signal in the region of L1. Instead, it transmits in the S band (2492 MHz) and also coincident with GPS L5 (1176.45 MHz).

Chapter 8. Airspace Management

1 Introduction

From a user perspective, not all pieces of airspace are equally important. Certain aircraft operators will have more interest in overflying some areas rather than others. This is at the very core of Airspace Management and Airspace Design. We have to take into account the needs of the users when defining available routes as well as the services to be provided.

From an ANSP perspective, efforts are allocated in those areas where demand requires. This means adapting CNS capabilities (communication and radar antennas, as well as ground navaids) to maintain minimum standards, seeking for flight safety and efficiency. Managing traffic safely requires adherence to certain rules or to the instructions given by air traffic controllers (when delegating the responsability for self-separation is considered unfeasible). It is easy to understand that airspace at lower levels in the middle of poorly populated areas does not require the same level of attention than the vicinity of a busy airport.

ICAO defines airspace management as the process by which airspace options are selected and applied to meet the needs of the airspace users.

2 Airspace volumes

These definitions are in accordance to ICAO Annex 2, Rules of the Air. Aiming at defining different categories, liabilities and services to be provided within each of them, the following airspace volumes and areas are defined.

- Landing Area - That part of a movement area intended for the landing or take off of aircraft. Typically runway(s).

- Maneuvering Area - That part of an aerodrome to be used for the take off, landing and taxiing of aircraft, excluding aprons. Typically runway(s) and taxiway(s).

- Movement Area - That part of an aerodrome to be used for the take off, landing and taxiing of aircraft, consisting of the maneuvering area and the aprons. Typically runway(s), taxiway(s) and apron(s).

- Aerodrome Traffic Zone (ATZ) - Airspace of defined dimensions established around an aerodrome for the protection of aerodrome traffic. Typically ATZs are cylinders centered on the ARP (Airport Reference Point) and have a radius of 5 nautical miles, with a height of 2500ft.

- Control Zone (CTR) - A controlled airspace extending upwards from the surface of the Earth to a specified upper limit. CTRs are designed to protect the approach maneuvers, embracing ATZs and serving as connections between ATZs and CTAs.

- Control Area (CTA) - A controlled airspace extending upwards from a specified limit above the Earth, to be at least 700ft above ground or water. They shall be delineated taking into account the capabilities of the navigation aids normally used in the area. Lower limits of Control Areas normally coincide with VFR cruising levels (which are flight levels that instead of ending in 0, as for the IFR, end in 5, for example, FL115).

- Terminal Manoeuvring Area (TMA) - A control area normally established at the confluence of ATS routes in the vicinity of one or more major aerodromes.

- Flight Information Region (FIR) - An airspace of defined dimensions within which flight information and alerting services are provided[1]. Flight information regions shall be delineated to cover the whole of the air route structure to be served by such regions.

2.1 ATS routes. Structured vs Non-Structured Airspace

Depending on whether or not aircraft are forced to follow certain predefined paths (typically ATS routes, with the shape of a corridor), we can find structured or non-structured airspaces. The complexity of the system (a significant amount of aircraft with changing altitudes and speeds, evolving through different tracks and with different intentions) is reduced forcing them to follow predefined routes (structured airspace), although a more efficient navigation (and also more complex) can be pursued in non-structured airspaces, where a free trajectory choice is permitted. This is obviously related with the Free Route concept and is currently being (halfway) implemented in some areas over Europe.

ATS routes are specified routes designed for channelling the flow of traffic as necessary for the provision of air traffic services. The term "ATS route" is

---

[1]UIR stands for Upper Information Region and it is a Flight Information Region in upper airspace (typically above FL245)

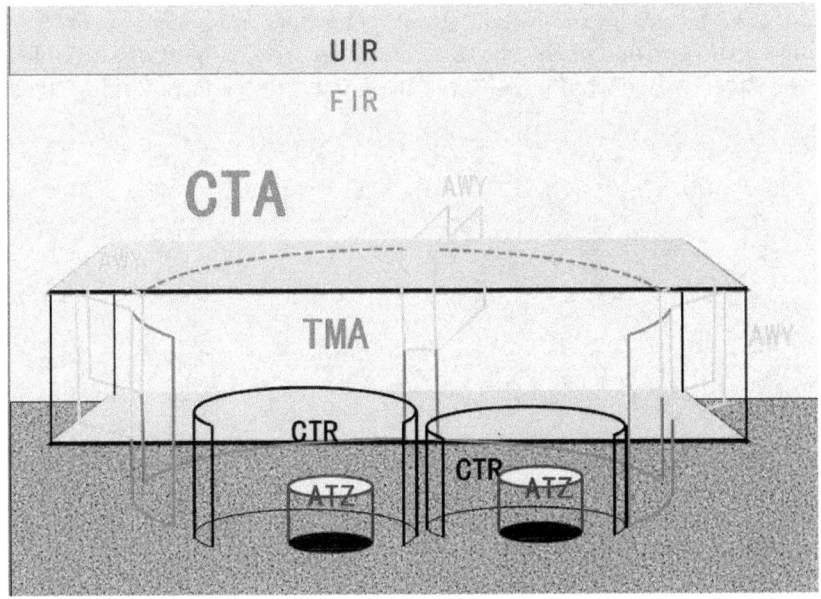

Figure 8.1: Example of typical airspace volumes distribution © Flazala / Wikipedia / CC BY-SA 3.0

used to mean variously, airway, advisory route, controlled or uncontrolled route, arrival or departure route, etc. An ATS route is defined by route specifications which include an ATS route designator, the track to or from significant points (way-points), distance between significant points, reporting requirements and, as determined by the appropiate ATS authority, the lowest safe altitude.

ATS routes are thus specified by each State, that is responsible for organizing the different traffic flows safely and efficiently. States must coordinate with each other how routes are connected through waypoints in the boundaries of the different airspace volumes when assigned to different ATS units. Two different networks of ATS routes are defined, one for lower airspace (FIR) and another one for upper airspace (UIR), being lower airways typically allocated to allow ground navaid based navigation (VOR or NDB based) and upper airways to be flown with RNAV equipment (more efficiently, not relying on ground navaids for the determination of the different way-points). States may also define a set of ATS routes to be used during the night (when demand and military activity is expected to be so low that more direct routing can be permitted).

The efficiency of a route network comes determined by the availability to cross certain areas in the most direct way. This efficiency is thus affected by the presence of restricted or military areas where overflight is not permitted (and that have to be avoided). There are also cases in which ATS routes are defined differently every day, as is the case of the North Atlantic Oceanic Track System (OTS), where depending on the position of the North Atlantic jet stream, a set of available routes are defined differently each day (way-points defined every ten meridian degrees).

3 Airspace classification

Since not all airspace volumes are equally important, different classes are defined. States shall define airspace classification according to their needs. Authorities may also define certain volumes as RMZ (Radio Mandatory Zones) or TMZ (Transponder Mandatory Zones).

Class	Type of flight	Separation provided	Service provided	Speed limitations	Radio required	Entry subject to ATC clearance
A	Only IFR	All aircraft	Control	N/A	Yes	Yes
B	IFR	All aircraft	Control	N/A	Ye	Yes
	VFR					
C	IFR	IFR from IFR & VFR	Control	N/A	Yes	Yes
	VFR	VFR from IFR	Control (vs IFR) & Information/Advice (vs VFR)	250 KIAS below 10000ft AMSL.	Yes	Yes
D	IFR	IFR from IFR	Control (vs IFR) & Information/Advice (vs VFR)	250 KIAS below 10000ft AMSL.	Yes	Yes
	VFR	No	Information/Advice (vs IFR & VFR)	250 KIAS below 10000ft AMSL.	Yes	Yes
E	IFR	IFR from IFR	Control (vs IFR) & Information (vs VFR)	250 KIAS below 10000ft AMSL.	Yes	Yes
	VFR	No	Information (vs IFR & VFR)	250 KIAS below 10000ft AMSL.	No	No
F	IFR	IFR from IFR if possible	Information/Advice (vs IFR & VFR)	250 KIAS below 10000ft AMSL.	No	No
	VFR	No	Information (vs IFR & VFR) if requested	250 KIAS below 10000ft AMSL.	No	No
G	IFR	No	Information (vs IFR & VFR) if requested	250 KIAS below 10000ft AMSL.	No	No
	VFR					

Table 8.1: Airspace classification

Controlled airspace is thus classified as class A, B, C, D or E, while uncontrolled airspace is classified as class F or G. Where the ATS airspaces adjoin vertically, that is, one above the other, flights at a common level would comply with requirements of, and be given services applicable to, the less restrictive class of airspace. In applying these criteria, Class B airspace is considered less restrictive than Class A airspace, Class C less restrictive than Class B airspace, etc.

4 Airspace restrictions

There are currently two main airspace users. On one side, civil operators (through private, commercial or state-owned aircraft), and on the other side, militar operators (transport, rescue and defence missions). The use they make of the airspace can sometimes be strongly different, making it impossible to integrate them under the same space, requiring a segregation of airspace. States have sovereignty over their airspace, and can establish a set of areas through which flight is not recommended, restricted or prohibited. These areas[2], when activated, do not have any assigned airspace class, and when deactivated have the airspace class of the surrounding airspace.

- Delta Area - Danger area. An airspace of defined dimensions within which activities dangerous to the flight of aircraft may exist at specified times.

- Papa Areas - Prohibited areas. An airspace of defined dimensions, above the land areas or territorial waters of a State, within which the flight of aircraft is prohibited.

- Romeo Areas - Restricted areas. An airspace of defined dimensions, above the land areas or territorial waters of a State, within which the flight of aircraft is restricted in accordance with certain specified conditions.

LED50 BARDENAS (Zaragoza)

420200N 0012200W; 421100N 0011000W;
423130N 0013900W; 422000N 0014800W;
420200N 0012200W.

FL 235
SFC

Ejercicios de tiro aire-tierra y bombardeo.
Air-ground firing and bombing exercises.

MON/FRI: 0700-1530
Otra actividad anunciada por / Other activity
announced by NOTAM. Coordinación con /
Coordination with: ZARAGOZA APP.

**LEP162 PALACIO REAL Y CAMPO DEL MORO
(Madrid)**

Círculo de 0,5 NM de radio con centro en / Circle of
0.5 NM radius centered on: 402503N 0034300W

4000 ft ALT
SFC

Prohibido el sobrevuelo.
Overflying is prohibited.
Permanente / Permanent.

LER93 MARISMA DE HUELVA

371642N 0065930W; 371548N 0065700W;
371054N 0065730W; 371242N 0070118W;
371642N 0065930W.

6000 ft ALT
SFC

Zona Ecológica. Protección y conservación de
las aves / Ecologic area. Migratory birds
protection.
Permanente / Permanent.

Figure 8.2: Example of D, P and R areas as published in the Spanish AIP

Some of these areas might be permanent, have a published schedule or are activated via NOTAM publication (typically for military exercises). ICAO Annex

---

[2]In Figure [8.2] notice that each area type is preceded by "LE", according to ICAO nomenclature for the Spanish Peninsula

2 establishes that aircraft shall not fly in prohibited or restricted areas, being the particulars of which duly published (typically in the AIP of the State to which the area belongs), except in accordance with the conditions of the restrictions or by permission of the State over whose territory the areas are established.

For cases of war or when military activity is expected to be very intense, States may define No-Fly zones, areas of airspace where aircraft shall not enter by any means. Aircraft that break the No-Fly restriction may be shot down[3].

Temporary Segregated Areas (TSA) are defined as pieces of airspace temporarily reserved and allocated for the exclusive use of a specific user during a determined period of time (e.g. military activity over a non-standard area, areas reserved for UAV use, etc).

ICAO has published circular 330, regarding how Civil - Militar coordination in ATM has to be for an efficient use of airspace. The following definitions are set:

- Flexible Use of Airspace (FUA) - An airspace management concept based on the principle that airspace should not be designated purely as civil or military, but rather as a continuum in which all user requirements are accommodated to the greatest possible extent.

- Collaborative Decision Making (CDM) - The process whereby all ATM decisions, except tactical ATC decisions, are based on sharing of all information relevant to air traffic operation between all civil and military partners.

- Temporary Segregated Area (TSA) - An airspace temporarily segregated and allocated for the exclusive use of a particular user during a determined period of time and through which other traffic will not be allowed to transit.

- Temporary Reserved Area (TRA) - An airspace temporarily reserved and allocated for the specific use of a particular user during a determined period of time and through which other traffic may be allowed to transit under air traffic control (ATC) clearance.

States do explicitly normally recognize that their respective Air Forces are responsible for preserving the sovereignty of the airspace over their territories, and that it is only during peace times that they let civil ANSPs manage their national airspaces. Air Forces have their own needs in terms of airspace to conduct training exercises and maneuvers, and thus define a set of areas that can be activated for these purposes. However, as ICAO recognizes, strong collaboration is required for a more efficient use of the available resources, avoiding excessive penalties over civil aviation. In Circular 330, interoperability is encourged through the following lines that can be found in Chapter 2: *The price of strategic and/or political interoperability at national as well as international levels can be high and finding*

---

[3]A controversial and recent event concerning the shootdown of MH17, a B777 flying from Amsterdam to Kuala Lumpur, raises concerns about current safety levels when overflying areas of conflict

a common ground can be difficult to achieve. National considerations and culture are potential disablers of affordable interoperability. Nevertheless one can assume that the aviation chain is as strong as its weakest link and that it is therefore in everyone's interest to cooperate and invest in order to achieve the highest level of interoperability.

5 Flexible Use of Airspace. Conditional Routing. Route Availability Document

5.1 Flexible use of airspace (FUA)

The different airspace restrictions that we have seen in section 4 limit the availability of more direct routes (and associated shorter flight times) for civil aircraft operators. Taking into account that military areas are not used 100% of the time, ICAO recommends a flexible use of these areas, avoding permanent airspace segregation. Through close coordination between civil and military authorities (in the strategic phase) and civil and military air traffic controllers (in the tactical phase), a more efficient allocation of airspace resources can lead to important savings for the airlines[4]. This coordination, ICAO says, requires a collaboration based on communication, education, a shared relationship and trust. Civil and military authorities should put in place procedures to apply airspace reservations or restrictions only during limited periods of time, based on actual use.

In order to be able to succeed in the flexible use of airspace, interoperability between civil and militar systems is required. This means that the civil system should be able to develop the ATM/CNS concept within restricted airspaces when they are not activated, being also able to manage the different functions over military users. This apparently easy task has still today its difficulties because of different reasons such as budget constraints, lack of space in militar cockpits for extra avionics, lack of certification[5] (leading on many occassions to state exemptions for the military users), etc. Today, it is common to find military areas activated with no activity inside, showing proof of current inefficiency in the system. Not to say that international cooperation in terms of FUA is still a pending subject. States show strong reluctancy when sharing what they consider sensitive information. Collaboration between international ANSPs and military authorities from different countries is (in most cases) inexistent despite of ICAO recommendations.

5.2 Conditional routing (CDR)

Within the FUA concept, ICAO proposes the use of Conditional Routes, defined as non-permanent ATS routes or portions thereof which can be planned and used

---

[4]Civil operators would not only fly shorter distances, but would also be able to fill in shorter Flight Plan distances, loading up less fuel and reducing the associated cost to carry

[5]Fighter jets, for instance, are normally not RVSM certified

under specified conditions. According to its foreseen availability, flight planning possibilities[6] and the expected level of activity of associated affecting areas, conditional routes can be divided into the following categories:

- Category One: permanently plannable. Aircraft Operators (AOs) can fill them in their respective flight plans, and if they are finally not usable, AOs are notified by the Network Manager before the operation takes place, being thus forced to change their Flight Plan.

- Category Two: non-permanently plannable. AOs can not fill them in their respective Flight Plans unless they are notified by the Network Manager about the availability of the route (typically not before the day of the operation).

- Category Three: not plannable at all. These CDRs can only be used upon coordination between civil and militar ATCOs in the tactical phase.

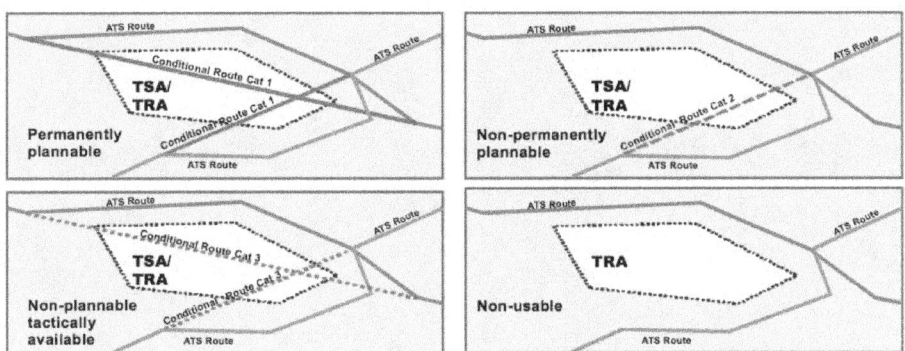

Figure 8.3: CDR classification © ICAO Circular 330. *Reproduced with the permission of ICAO*

5.3 Route Availability Document (RAD)

The RAD is a document published by Eurocontrol, helped by different ANSPs and State authorities in which certain restrictions are applied to the ATS network. For ATFM reasons mostly, restrictions are applied to certain routes or flight plans from a strategic perspective. These restrictions can be:

---

[6]Be aware that flight planning determines fuel uplift. Whether or not a certain ATS route is going to be available is thus essential for fuel computations and may affect the safety of the flight.

- Maximum flight levels for certain routes (to reduce complexity in the network, specially for short routes in which aircraft may tend to fly parabolic profiles) - avoiding upper sector congestion.

- Lateral predefined reroutings - to avoid certain airspace sectors, being affected grouped flight plans (typically selected depending on their origin) with the same destination.

- Time limitations to certain ATS routes - some routes might only be available during a published schedule.

Chapter 9. Air Traffic Flow Management

ICAO defines ATFM as a service established with the objective of contributing to a safe, orderly and expeditious flow of air traffic by ensuring that ATC capacity is utilized to the maximum extent possible and that the traffic volume is compatible with the capacities declared by the appropriate ATS authority.

1 Flight Plan within the ATFM service

As we have already seen through different chapters and we will more specifically cover when we talk about ATS Services, aircraft require some services to be provided by stakeholders other than the aircraft operator (i.e. airports and ANSPs). Moreover, aircraft do not fly alone and the configuration of the airspace, specifically for the provision of control services, requires to know accurately the expected amount of traffic and the intentions of the different aircraft. A Flight Plan is required as a document through which each aircraft reports its flight intentions. ANSPs need to know which are the requirements of each flight in terms of lateral and vertical navigation (intended ATS routes and requested flight levels, for each instant of the flight) and what is the performance of the aircraft they are handling (what are they capable or not capable of doing), while other services (e.g. alert services) may require more precise information about the number of people onboard in case a search and rescue operation has to be performed. The need for a document to be submitted prior to the departure of the flight, containing all this information, becomes thus evident.

ICAO Annex 2 defines the following flight plan versions:

- Flight Plan (FPL) - Specified information provided to air traffic services units, relative to an intended flight or portion of a flight of an aircraft.

- Filed Flight Plan - The flight plan as filed with an ATS unit by the pilot or a designated representative, without any subsequent changes.

- Repetitive Flight Plan (RPL)[1] - A flight plan related to a series of frequently recurring, regularly operated individual flights with identical basic features, submitted by an operator for retention and repetitive use by ATS units.

- Current Flight Plan[2] - The flight plan, including changes, if any, brought about by subsequent clearances.

ICAO establishes which flights must compulsorily present a Flight Plan prior to the departure of the aircraft:

1. Any flight that will require air traffic control services

2. Any IFR flight expecting to fly within airspace classes different than G

3. Any flight requiring information, alert or search and rescue services

4. Any flight across international borders

Pilots are responsible for the presentation of the Flight Plan (although currently most airlines have Flight Dispatchers helping the fulfillment of Flight Plans, requiring just the approval of the captain). They have to be presented to the ATS Reporting Office (ARO) associated to the departure aerodrome (that later, through a ground-based network, communicates with the different ATS Units, providing them with the necessary information). However, if ARO offices are not available, and typically for the case of remote aerodromes and VFR operations, it is permitted to fulfill the Flight Plan once airborne through the radio (communicating as soon as possible with the assigned ATS Unit), this is called AFIL (Air-FILed flight plan).

The content of the Flight Plan comes regulated by ICAO Annex 2 Chapter 3, and has to include to following information:

---

[1]Typically, RPLs do not fit real operational requirements in the tactical phase, since they are filed months in advance, not knowing accurately which will be magnitudes suchs as the weight of the aircraft (determined by the number of passengers), wind profiles, temperature distribution of the atmosphere, etc.

[2]In Spanish we name it *Plan de Vuelo Actualizado* and only makes sense during the tactical phase

Concept	Example
Aircraf Identification (Callsign)	IBE04UV
Flight Rules (*VFR, IFR, Z or Y*) and Type of Flight (*Scheduled air service, Non-scheduled air service, General aviation, Militar, etc*)	IFR - Scheduled air service
Number of aircraft (*Typically for military formation operations*), aircraft type and wake turbulence category	A320 - Medium
Equipment (*A set of codes are defined in ICAO Doc 4444 for each aircraft equipment*)	8,33kHz equipped, RVSM and PBN approved
Departure Aerodrome	LEMD
Estimated Off-Block Time (EOBT) - *Always using GMT times*	0935
Cruising speed - *True Airspeed given in knots, or cruising Mach Number*	M078
Cruising Level	F280
Route to be followed - *From the last point of the intended SID, to the first point of the intended STAR*	PINAR UN870 PONEN UT600 CASPE
Destination Aerodrome and Total estimated elapsed time	LEBL 0044
Alternate Aerodrome - *An aerodrome to which an aircraft may proceed when it becomes either impossible or inadvisable to proceed to or to land at the aerodrome of intended landing where the necessary services and facilities are available, where aircraft performance requirements can be met and which is operational at the expected time of use*	LEVC
Fuel endurance	02H00M
Number of persons onboard	157

Since the activation of a Flight Plan means that there are third parties encharged of surveiling and paying attention to the safe completion of the flight, the lack of an aircraft arrival report will trigger alert services. To avoid unnecessary alert activations, it is important to properly close the Flight Plan once the flight has been successfully completed.

Flight Plans are used by different actors during all phases (strategic, pretactical and tactical).

- In Europe, ARO Offices send the Flight Plans they collect to the IFPS (Integrated Initial Flight Plan Processing System) in Brussels, that provides information to the Network Manager in order to foresee the matching between demand and capacity. Global demand is predicted through the use of individual Flight Plan information.

- Flight Dispatchers and Flight Crew use Flight Plan information to compute the amount of required fuel uplift (depending on expected lateral and vertical routes, winds, weights, etc).

- Flight Crew uses Flight Plan lateral intentions to enter the route into the onboard Flight Management System (FMS).

- Air Traffic Controllers find Flight Plan information presented on their screens, which is basic for trajectory management and conflict avoidance, since it indicates the intentions of each aircraft.

Flight planning is the process of producing a Flight Plan, requiring accurate weather forecasts for the determination of optimum lateral and vertical route profiles. An effective Flight Plan can reduce fuel costs, time-based costs, overflight costs[3] and lost revenue from payload that can not be carried (because of weight limitations) simply by efficiently modifying the lateral route, vertical altitudes and speeds.

2 ATFM structure and metrics

The main purpose of ATFM[4] is to avoid sector and airport overloading, by controlling the match between capacity and demand. In order to do so, a set of executive bodies have been established, a network to transmit information, as well as ways to measure demand and capacity. The NMOC (Network Manager Operations Center) is the core of this system, that monitors demand vs capacity matching and assigns regulations when needed with the help of FMPs (Flow Management Positions), typically allocated at each ACC (Area Control Center), that exchange information with the different sectors and airport towers under their jurisdiction.

Since ATFM relies mostly on forecasts, it is necessary to understand the three mainly defined phases, where different actions and accuracy of predictions can be expected. However, they must not be understood as discrete phases, instead, they act continuously, adapting flexibly to the conditions of the environment:

- Strategic Phase - From months before the operation to 48 hours before initially filed EOBT.

- Pretactical Phase - From 48 hours before the EOBT to a few hours before the operation.

---

[3]LIDO commercial software (developed by Lufthansa) already takes into account ANSPs charges to reduce the total cost of the flight

[4]The following lines will express the way Flow Management is performed in Europe, since it is the daily environment that the author knows professionally

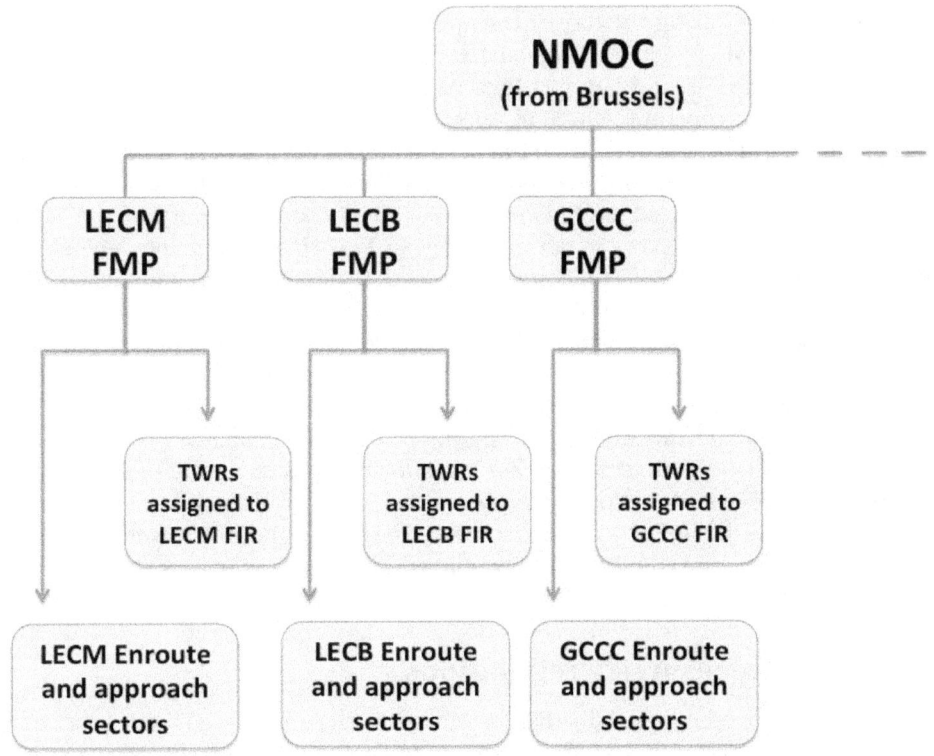

Figure 9.1: Structure of ATFM network in the Eurocontrol area, exemplified with Spanish FMPs

- Tactical Phase - From a few hours before the operation to the end of the flight.

2.1 Demand and capacity

On the side of the **demand**, we can say it is typically accounted for by the number of movements. Within an airspace sector, two basic metrics are defined (see Figure [9.3]), HEC (Hourly Entry Counts, that counts the number of aircraft that will enter the sector every hour) and OCC (Occupancy Counts, that counts the number of aircraft that will be within the boundaries of an airspace sector at each given instant). In an airport environment, traditional metrics count the number of movements per runway (take-offs plus landings). It is with the information provided by aircraft operators via Flight Plans that the demand is estimated, taking into account the expected flight levels (to see which upper or lower sectors the aircraft is going to fly through) and speeds (to compute entry and exit times). Any disturbance occurring during the day of the operation (late departure of the aircraft, stronger winds than expected, assigned FL different from FPL, etc) leads to lack of adherence to the Flight Plan and thus to real demand not matching expected demand, which is a constant source of uncertainty for the ATFM system.

Capacity is typically determined as the maximum number of movements an ATS Unit (airport or airspace sector) can handle in terms of HEC or OCC. It is a threshold value over which it is considered that the complexity of the ATM situation is so high that safety is not guaranteed. This threshold value depends on many different factors (which makes capacity estimation strongly complex, as can be checked in Figure [9.4]). Among others, capacity will depend on the complexity of the network of ATS routes within the sector (whether or not there are crossings and how complex these crossings are), the expected performance of the aircraft within the sector (some aircraft types, like fighter jets, happen to maneuver very quickly, increasing the uncertainty of the flight variables displayed on the radar screen), the presence of adverse weather phenomena (presence of CBs sometimes leads to a decrease in sector capacity because it causes aircraft deviations and limits the amount of available FLs) and of course, controller equipment and workload (e.g. CPDLC reducing radio communication workload, or radar versus procedural air traffic control).

2.2 ATFM measures - regulations

If for any of the reasons explained in section 2.3 there is a predicted mismatch between capacity and demand that can be foreseen with enough anticipation, measures are taken to avoid sector overloading. If these overloadings happen repeatedly in time, a strategic level action can be expected (modification of ATS Routes, sector distribution, staff resources, ATC procedures or RAD). However, most ATFM measures happen in the pretactical and tactical phases and have to do with a modified trajectory, in terms of lateral, vertical or time modification of flight intentions with respect to the initially filed Flight Plan. These are named "**regulations**" and the most common are:

- **Ground delays** - demand curves (HEC & OCC) are artificially moved towards the future by assigning ground delays to certain flights, until capacity matches expected demand. In Europe, each flight has a Departure Tolerance Window (DTW), which is a time window within which a flight has to depart according to his flight plan. The NMOC estimates the time at which an aircraft is going to depart with the following formula: ETOT = EOBT + Taxitime. Where ETOT stands for Estimated Take Off Time and Taxitime is a fixed value for the departure airport (that can be modified tactically by the tower supervisor if necessary). Under normal operations (no regulation assigned), the NMOC will assume a tolerance window for a non-regulated flight, being DTW = [ETOT-15', ETOT+15']. If the aircraft, for any reason, can not comply with this tolerance window, must then update his flight plan. When regulations are enforced, certain aircraft are assigned a CTOT (Calculated Take Off Time), which is equal to EOBT + Taxitime + Enforced Delay (by the NMOC). DTW when CTOT is assigned is [CTOT-5', CTOT+10']. This information is shared among the different agents through networks named AFTN (tower) and ACARS (cockpit) with messages such as SAM (Slot Allocation Message), SRM (Slot Revision Message), REA (Ready Message), etc.

- **Level Capping** - because of overflights provoking congestion in upper airspace sectors, some departures may be limited in their climb, remaining in lower sectors until peak levels of traffic are reduced. These regulations are typically not transmitted to the flight crew before the departure, and instead are taken during the tactical phase. When upper sectors start to be relieved, one by one coordination can be made to allow more aircraft to climb to their preferred level. This action is normally triggered by ATC supervisors assisted by FMP personnel.

- **Rerouting** - because of the existence of capacity limits in some sectors, flight dispatchers, in close coordination with the NMOC may propose lateral or vertical rerouting solutions to avoid the regulated areas.

- **Rate restrictions** - when unforeseen events take place (such as unexpected storms, wild fires in the vicinity of an aerodrome, etc), or a desirable regulation policy has not been put into practice well in advance, the ATM situation can deteriorate to levels at which only tactical action can be taken. If this happens, rate restrictions are coordinated between different ACCs or even sectors within the same ACC (sectors not accepting traffic or accepting a limited amount of traffic). The hardest of all restrictions is named "Rate Zero" and means that a given ATS Unit accepts no traffic at all. The absence of CTOT under these circumstances makes pilots not understand very well why their departure is delayed by the tower. ATFM reports such as the ones Eurocontrol publishes every year do not take into account these circumstances since they happen so downstream in the production chain that no regulation is put into paper (CTOTs are not assigned).

Any taken measure will have its own operational consequence (from a delayed ground startup time, to a restricted assigned FL). These are always applied by the last part of the chain (basically air traffic controllers that force pilots to comply with certain restrictions). These tactical operators should have the necessary information to understand and execute measures that make sense from a global perspective (although maybe not locally). This is key when talking about ATFM, each agent must understand that the actions taken at a given time and position will have an impact later on at some other point of the network (downstream). Lack of flight plan adherence, introduced noise and uncertainty in the network make the

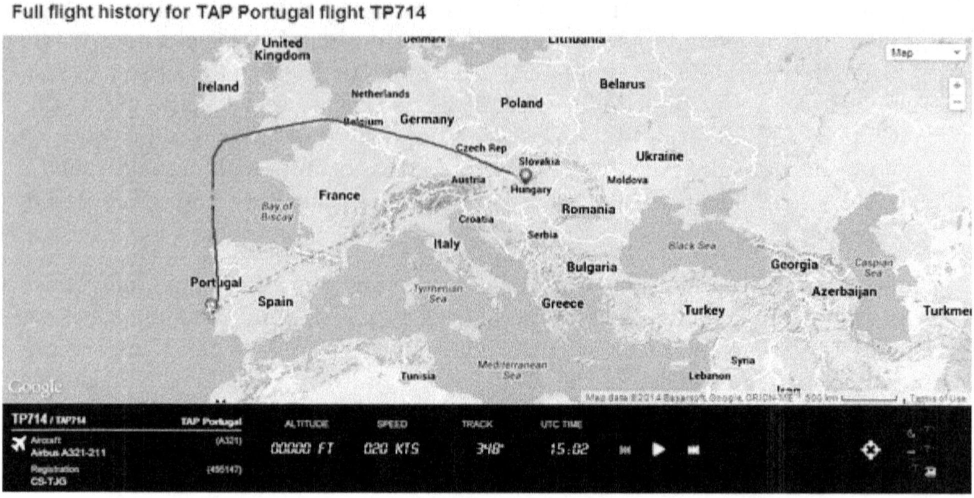

Full flight history for TAP Portugal flight TP714

Figure 9.2: Rerouting example of TAP117 avoiding French Airspace because of very strong regulations © FR24

ATFM system fail in its predictions, provoking sometimes an unbalanced situation between capacity and demand (compromising sometimes safety and other times efficiency).

2.3 Causes of the regulations

Every time a CTOT is assigned (ground delay), the reason associated to the regulation is attached in the message (properly coded). The most typical causes[5] are:

- **ATC Capacity** - demand simply overpassing capacity of a given ATC sector

- **Aerodrome Capacity** - airports sell their parking slots to the airlines well in advance before the operation. If delays during the day accumulate aircraft on a given aerodrome (true demand not matching pretactical expected demand), there might not be enough resources to accommodate the initially sold parking slots, imposing ground delays at their origins (i.e. lack of parking space, different restrictions associated to safety agents such as firemen, etc).

- **ATC Staff** - lack of ATC personnel leading to a lower number of open sectors or ATC positions that can not handle as much traffic as expected.

---

[5]Take into account that here we are not talking about causes of delay in general (e.g. maintenance problem, passenger not showing up with luggage onboard, etc). We are only dealing with ATFM causes

- **Weather** - presence of CBs increase the uncertainty of aircraft trajectory, boosting complexity, communications and controller workload. The number of aircraf that can be handled is lower, and thus capacity is reduced.

- **System failure** - radar, radio or datalink failures imply restrictions to the capacity of a given area (the number of aircraft that can be handled is lower)[6].

- **Special events** - flight exhibitions or other airspace restrictions because of any special event may lead to ATFM regulations.

---

[6]Examples of this can be the closure of LEMG in April 23rd 2015 due to radar failure or the UK computer meltdown that caused a rate zero during December 12th 2014

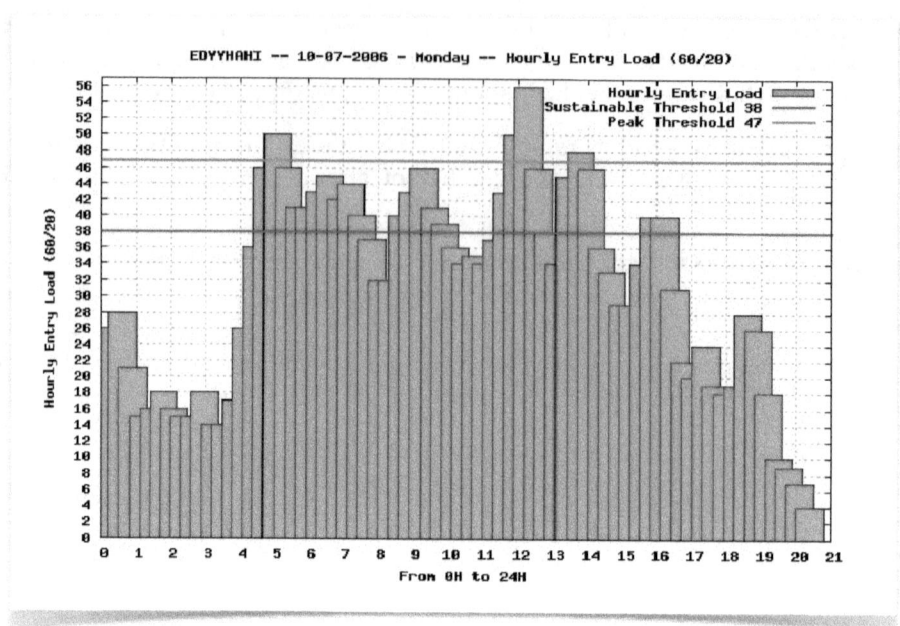

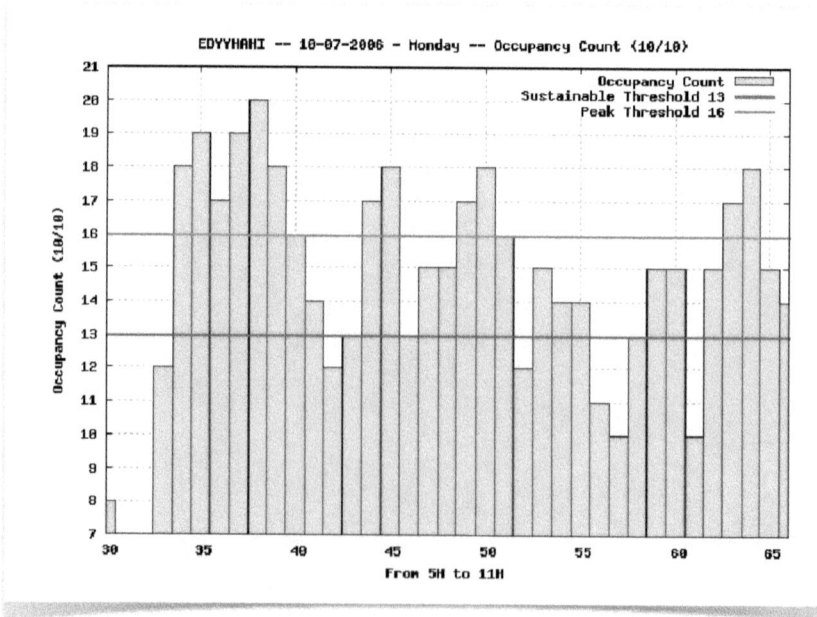

Figure 9.3: Basic plots showing prediction of HEC & OCC for a given airspace sector © Eurocontrol

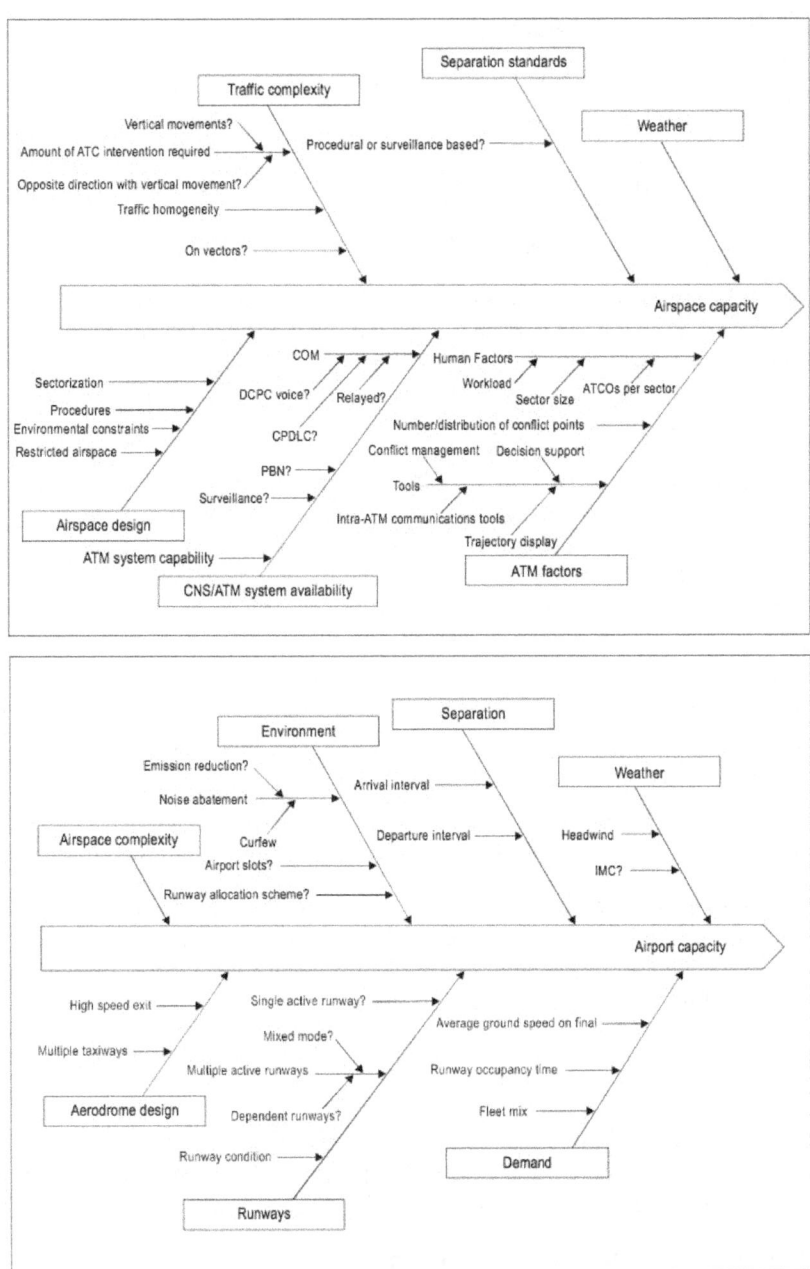

Figure 9.4: Factors that affect airspace and airport capacity as expressed in ICAO Doc 9971 © ICAO Doc 9971. *Reproduced with the permission of ICAO*

3 ATFM uncertainty

As we have seen in section 2.2, different actions can be taken in order to try to mitigate a predicted lack of balance between capacity and demand. However, on many occassions the presence of uncertainty in the system makes basic variables such as Times Over and Flight Levels be different from forecasted values.

Demand and capacity predicted values may not match real situation evolution because of different factors, among others:

- **Lack of adherence to the established Departure Tolerance Window** - when tower ATCOs do not comply with established procedures and pilots accept a departure outside the DTW margins. This lack of adherence causes a change in all the time over estimates for the different entry and exit points of the affected airspace sectors.

- **Lack of adherence to Flight Plan requested flight levels** - because of dynamic evolution of aircraft weight, winds aloft and temperature profiles in the atmosphere, forecasts made by Flight Dispatchers[7] may not match the most optimum solution for the aircraft, requiring later the pilot, in the tactical phase, a deviation from the FPL seeking for more optimum vertical profiles. ATCOs may change the FL of an aircraft within a given sector if no traffic is affecting the aircraft, not taking into account the consequences this may have downstream in the system.

- **Winds** - from airport configuration to enroute Ground Speeds, wind continuously determines how the system performs (in terms of trajectories and times). A bad forecast regarding expected winds will affect expected speeds and thus expected entry and exit times in the different airspace sectors.

- **Formation of CBs** - regarding capacity restrictions, information about the formation of CBs (position at which they will develop and timing) is usually not accurate enough, not being able to properly predict (with high accuracy) what will be the impact on the operations.

When predicted values of demand and capacity do not match reality in the end, there can be two different situations:

- **Real capacity exceeding real demand** - if real demand is lower than predicted, there might be a buffer of capacity in the system not being used, while penalized aircraft have been regulated. Efficiency is compromised.

- **Real demand exceeding real capacity** - safety and efficiency can be compromised because of sector overloading, leading to:

---

[7]Specially for the case of RPLs

- Congestion in TMA areas - holding patterns imposed to aircraft, that have to wait holding over different points wasting a lot more fuel than if they had waited on the ground (at their departure airport)

- Congestion on taxiways and aprons at the airport of arrival - arriving aircraft may not find parking space because of apron saturation. This may lead to delays on the ground with the engines running, not being the passengers able to leave the aircraft.

- Airprox events - losses of separation between aircraft because of increased levels of complexity, workload and airspace congestion.

Chapter 10. ATS Services

ATS Services are those provided by the ATS Units to aircraft during the tactical phase.

1 Objectives of Air Traffic Services

According to ICAO Annex 11, Chapter 2, the objectives of the air traffic services shall be to:

- Prevent collisions between aircraft when airborne and on the maneuvering area

- Prevent collisions between aircraft and terrain or obstacles on the ground

- Expedite and maintain an orderly flow of air traffic

- Provide advice and information useful for the safe and efficient conduct of flights

- Notify appropiate organizations regarding aircraft in need of search and rescue aid, and assist such organizations as required

2 ATS Units

ATS Units refer to the ground infrastructure from where ATS Services are provided. Basically we talk about Control Towers (TWRs) and Area Control Centers (ACCs). Control Towers provide service in the vicinity of the associated aerodrome, particularly to aircraft on the ground, the runway or during take off or landing, while Area Control Centers cover wider airspace areas and focus on the enroute and the approach flight phases. Figure [10.1] shows two typical ATS Unit dispositions.

Figure 10.1: Control Tower (left) and Area Control Center (right)

3 Classification of different ATS Services

3.1 Information Service

The information to be transmitted to the different aircraft as an ATS Service should not be mistaken with the information provided through the Aeronautical Information Service (AIS). While the AIS compiles a set of procedures, routes and relevant information that can be checked by different agents prior to the flight (e.g. for flight planning) or during the flight (e.g. navigation charts), the Information ATS Service consists of relevant tactical information that can be of the interest of the pilot to safely conduct the flight.

Air Traffic Controllers provide Information Service to all aircraft within their volumes of responsibility when suitable[1]. ICAO Annex 11, Chapter 4 gathers a set of standard practices about the information to be transmitted, being the most relevant from the operational point of view:

- Essential traffic information - within certain airspace volumes, where ATC can not instruct control clearances to provide air traffic separation (e.g. VFR vs IFR in D-class airspace), information regarding the relative position of the other traffic is transmitted expecting the pilots to achieve visual contact and avoid any sort of trajectory incursion.

- Significant weather information[2] - typically:

  - when enroute, the presence of ice, turbulence or CBs (that may make pilots request a flight level change or a lateral deviation).

---

[1]See airspace classification in the Airspace Management Chapter

[2]Part of this information is typically radiated using ATIS (Automatic Terminal Information Service) or VOLMET systems (with specific VHF frequencies)

- in the vicinity of an aerodrome, presence of windshear, wind direction and intensity, QNH, etc.

- Any sort of affection in the ground infrastructure that may compromise the safety of the flight. For instance:

  - navaids temporarily out of service (NDB, VOR, ILS)
  - runway status affecting aircraft braking action (wet, dry, presence of snow, etc).

- Any relevant information that may be considered of interest for the pilot - for instance:

  - Runway in use
  - Rapid exits of a runway blocked by other aircraft or airport working personnel
  - Presence of flocks of birds in the vicinity of an airport

Some uncontrolled aerodromes facilitate an Information Service named AFIS (Aerodrome Flight Information Service), where the personnel in the tower (who are not air traffic controllers, since they can not emit clearances or orders) give information to the pilots regarding operational status of equipment, weather information, positions of other traffic, etc. Under an AFIS environment, it is always the captain´s responsibility to maneuver in accordance with the Rules of the Air to avoid potential collisions. Some airports have also "apron towers" from where non-ATC personnel provide Information Services (over the apron, which is not part of the Maneuvering Area[3]).

3.2 Alert Service

After the activation of a Flight Plan (departure of the flight), the alert service should provide constant tracking of an aircraft (with a feasible resolution in time) to activate, if needed, search and rescue actions. In order to facilitate aircraft tracking in case of fatal accident, aircraft have to be equipped with an onboard ELT (Emergency Locator Transmitter), that in case of crash emits a signal in the 121,5 MHz frequency (this frequency is named *emergency frequency* and must be constantly monitored by all aircraft and ATS units secondary radio equipments).

According to ICAO Annex 11, Chapter 5, Area Control Centres (ACCs) shall serve as the central point for collecting all information relevant to a state of emergency of an aircraft operating within the concerned FIR and for forwarding such information to the appropriate rescue coordination centre. In the event of a state of emergency arising to an aircraft while it is under the control of an aerodrome

---

[3] As we already saw in the Airspace Management Chapter, section 2

control tower or approach control unit, such unit shall notify immediately the ACC responsible which shall in turn notify the rescue coordination center.

When providing the alert service, and when a certain aircraft is considered to be in trouble, the following phases are defined:

- **Uncertainty Phase (INCERFA)** - This phase is activated when no news are received from an aircraft 30 minutes after a transmission should have been received, or 30 minutes after the first unsuccessful attempt to communicate with such aircraft is made, or when an aircraft fails to land within 30 minutes after its last estimated landing time (whatever happens first).

- **Alert Phase (ALERFA)** - It is activated when, after INCERFA, several attempts to establish communication with the aircraft fail or when the aircraft fails to land five minutes after having received clearance to land (and communication is not re-established) or when the aircraft is known or believed to be the subject of unlawful interference.

- **Distress Phase (DETRESFA)** - Activated after ALERFA if continuous attempts to establish communication fail or there is evidence that the aircraft may have its airworthiness impaired and may have required an emergency landing, or when the fuel onboard is considered to be exhausted or insufficient for the safe conduct of the flight.

3.3   Advisory Service

Advisory Service is only provided in certain airspace classes (typically airspace class F) in phases prior to the implementation of Control Service. ICAO Doc 4444, in article 9.1.4 establishes that the Advisory Service objective is to make information on collision hazards more effective than it would be in the mere provision of flight information service.

It is considered to be a middle-point transition between Information and Control services for those areas where not full Control Service can be provided because of technical limitations (e.g. lack of radar or radio coverage, or sovereignty issues regarding the responsability of ATM provision within a given airspace). Air traffic advisory service does not afford the degree of safety and cannot assume the same responsibilities as air traffic control service in respect of the avoidance of collisions, since information regarding the disposition of traffic in the concerned area, made available to the ATS unit may be incomplete.

To make this quite clear, air traffic Advisory Service does not deliver "clearances" but only "advisory information" and it uses the word "advise" or "suggest" when a course of action is proposed to an aircraft. Air Traffic Controllers take no responsibility for separation when providing Advisory Service.

3.4 Control Service

Control service comes regulated by ICAO Annex 11, Chapter 3. It is to be provided inside the volumes of control that require it[4] to aircraft flying in accordance to the suitable flight rules (IFR or VFR depending on the airspace class). When providing control service, separation between different aircraft is provided by air traffic controllers (that assumme this responsability). In order to avoid losses of separation and keep the flow safe, air traffic controllers instruct "orders" to the pilots (known as "clearances"). A controlled flight shall be under the control of only one air traffic control unit at any given time. Each air traffic controller has his own piece of responsability (airspace sector or part of an airport) and technical means (surveillance + communication), with which he performs his tasks, preserving a safe, ordered and efficient use of the airspace. ICAO Doc 4444 establishes which can be the minimum separation requirements for each of the following environments.

- **Aerodrome Control Service** - This service is provided from a Control Tower (TWR). Within the tower there may be several air traffic controllers, each of them encharged of a given piece of airspace (or part of the airport) and with assigned tasks. Typically, there are three tower ATC positions:

  - **Clearance position** - the ATCO is encharged of issuing after departure route clearances (containing information about assigned SID, runway in use, initial stop altitude, QNH and transponder code) and issuing startup clearance times (checking DTW compliance). Aircraft are not in movement when under the control of this ATCO.

  - **GMC** - the Ground Movement Control ATCO is encharged of keeping aircraft that are taxiing separated from each other, assigning turns, establishing sequences and ordering aircraft to give way to each other. The ATCO normally relies on his visual contact with the different aircraft (checking their positions and supervising their movements), although in some advanced airports he may rely on a surface movement radar. An efficient GMC ATCO prepares an optimum departure sequence that can lead to optimum runway utilization (function of wake turbulence separation and others).

  - **LCL** - the Local ATCO is encharged of the use of the runway and the controlled flights within the Aerodrome Traffic Zone (ATZ). Surveillance is performed through visual observation of the aircraft sometimes supported by a radar screen where the departing and arriving aircraft are displayed. On the runway, the LCL ATCO has to make sure that only one aircraft is cleared to use the runway at the same time (he can not clear an aircraft for take off until the previous departure has lifted off the ground or the previous arrival has vacated the runway; and he

---

[4]See airspace classification

can not clear an aircraft to land until the preceding arrival has vacated the runway or the preceding departure has lifted off the runway). Depending on the number of runways and active configuration, airports may work under:

* **Segregated Runway Operations** - runways are used only for departures or only for arrivals. Typical metrics show capacities of the order of 30 movements per hour. LCL ATCO must take care, for consecutive departure operations, to comply with minimum wake turbulence separation (expressed in terms of time to elapse between the inducing and the trailing aircraft). He also has to comply with restrictions imposed by his downstream colateral unit for departures (the approach ATCO) that will typically require a minimum radar distance between two consecutive departures, as described in the Letter of Agreement between both units. Arriving aircraft are handed over to the LCL ATCO by the Approach Controller (typically established on final for IFR flights) and later cleared to land, however, if the runway were occupied by another aircraft or vehicle, he might have to instruct the arriving aircraft to perform a Go Around maneuver (Missed Approach).

* **Mixed Mode Runway Operations** - one single runway is used for both, departures and arrivals. It is the LCL ATCO job to determine whether an aircraft waiting at the holding point for departure can fit into two incoming arrivals. This intercalation job drives the efficiency of the runway and the uncertainty associated to it makes it critical for the safety of operations (loss of separation can lead to Rejected Take Off maneuvers or Missed Approaches). Figure [10.2] shows this mixed mode for runway operations.

LCL ATCO does also manage VFR traffic entering or exiting the aerodrome through the ATZ. He manages the different aircraft along the different circuit legs (see Figure [10.3]), clearing them to continue or instructing them to orbit in accordance with the expected sequence for the use of the runway[5].

- **Approach Control Service** - This service can be provided from a Control Tower (TWR), an Approach Unit (APP) or an Area Control Center (ACC). The APP ATCO is normally encharged of approach airspace sectors up to FL245, covering mostly CTRs and TMAs. The type of traffic that is handled within these levels makes these sectors the highest in terms of complexity, due to the constant evolution of speed, heading and altitude (aircraft climb

---

[5]It must be noticed that under the ATZ environment, safety nets such as TCAS do not work properly and are inhibited, as the density of traffic can be so high that it would be more a confusing rather than a helping system

Figure 10.2: Arriving aircraft behind the departing one already lifted off in Hong Kong - Kai Tak © Samuel Lo

up to their cruise level through a SID, or descend looking for the final leg of the approach through a STAR and a subsequent approach procedure). Besides traffic under evolution, cruise turboprops flying at low levels are also controlled by these units. The most challenging task for an APP ATCO is typically to organize and establish a sequence of incoming arrivals along the ILS of the runway in use, for this to be done successfully, constant monitoring and indications of speed, heading and altitude have to be given to the different aircraft, creating the sequence (see Figure [10.4]). Once the aircraft is established on final, it is transferred to the LCL ATCO in the tower. Airports with normal and large amounts of traffic require radar surveillance tracking for an efficient use of the airspace. Depending on the accuracy of the radar, wake turbulence categories and the Letter of Agreement between the APP and TWR ATS Units, the APP ATCO will have to provide a minimum separation value between two consecutive arrivals. Airports with lower traffic numbers might not have available radar surveillance and thus perform the approach control using "Procedural Techniques", assigning cleared altitudes to the new incoming aircraft, that typically have to hold over certain airspace fixes until separation is achieved. In the future, separations based on time instead of distance (as currently done when using the radar) expect to have important positive outcomes[6].

---

[6]A short video by NATS regarding TBS can be checked following the link

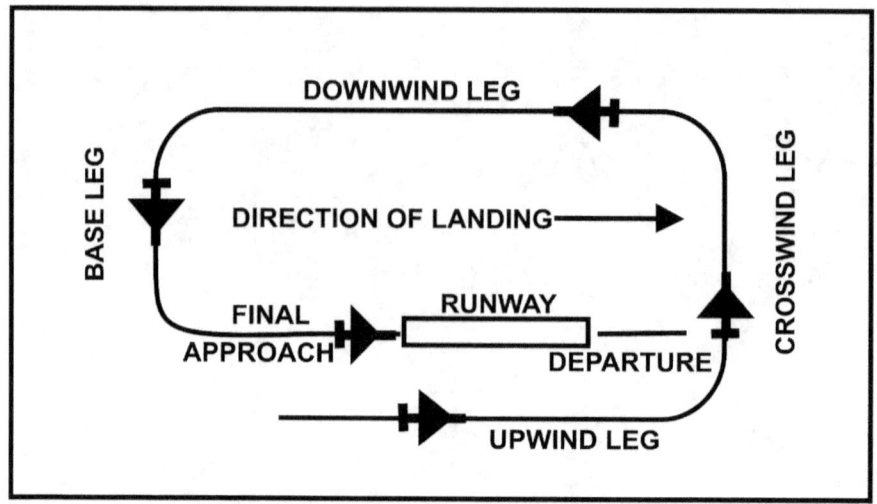

Figure 10.3: Aerodrome circuit for VFR traffic © Wikimedia / Public Domain

- **Enroute Control Service** - This service is provided from an Area Control Center (ACC), covering CTAs, FIRs and UIRs. One pair of ATCOs manages each airspace sector (depending on the size of the FIR and the expected demand, the number of sectors can be high, typically around 15-20 sectors). One ATCO works as the *executive controller* of the sector, giving the orders, handling the radio and speaking with the pilots, while the other one works as a *planning controller*, supporting his colleague to make the right decisions, coordinating with colateral units and assisting where/when needed. Sectors that have CTRs or TMAs underneath have to coordinate with approach ATS units. ACCs handle traffic during the last part of the climb, cruise and first part of the descent. ACCs typically organize the traffic avoiding conflicts (losses of separation) and assigning the proper Flight Levels to each aircraft (upon request, see Figure [10.5]). Conflict avoidance is typically pursued through aircraft vectoring (instructing the aircraft to turn right or left, increasing the number of miles for the crossing) or FL change. Figure [10.6] shows a pair of conflicting aircraft being deconflicted.

4 ATM Tactical Procedures

We talked about the different types of control service that can be provided in section 3.4, depending on the flight phase and associated ATS unit. Now we will

https://www.youtube.com/watch?v=2jBvbpbq0SM

130

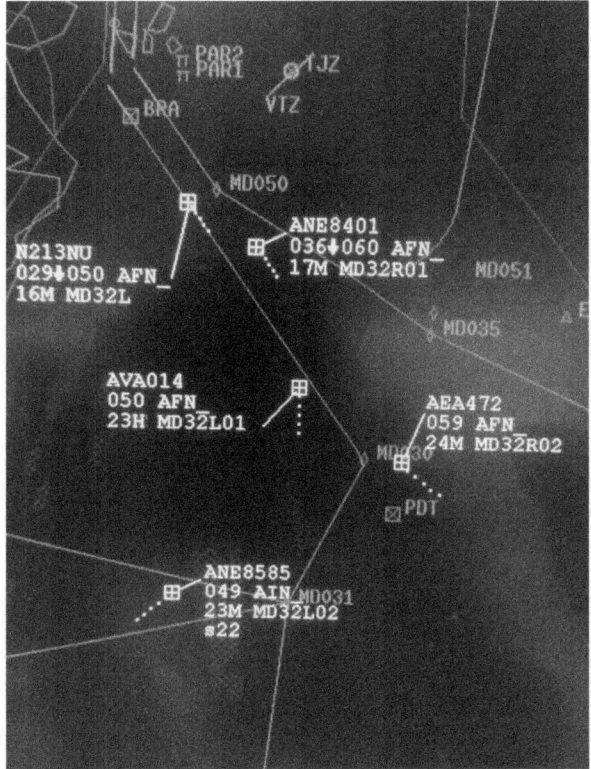

Figure 10.4: Sequence of arriving aircraft into Madrid Barajas runways 32L & 32R

get deeper into the ATM procedures that are applied during the tactical phase. What we summarize along the following lines are the basic tools and actions taken by air traffic controllers to manage the trajectories of the different airspace users safely in real time.

4.1   Separation standards

National authorities establish minimum separation standards to ensure the safe circulation of aircraft. A proper application of these standards facilitates a safe separation from the ground, from protected airspace volumes or from other aircraft. This means that aircraft cannot come closer than certain defined separation distances (or times) to the ground, other aircraft or some protected airspace volumes. Separation standards may also be enforced to reduce the exposure to wake vortex turbulence of any preceding aircraft.

In a 3D airspace environment, separation is defined differently in each direction:

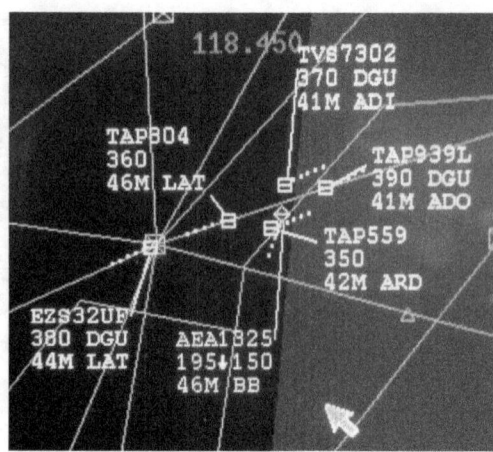

Figure 10.5: Busy ATS route with all RVSM Flight Levels from FL350 up to FL390 being used by different aircraft

- **Vertical separation** is required based on barometric measurements by different aircraft altimeters, as we already depicted in the Aircraft Instruments chapter. Depending on the enforced minimum vertical separation, we may be talking about a CVSM or a RVSM environment. Vertical separation below FL290 must always be at least of 1000 ft. Also, vertical separation must always be provided among aircraft placed on the same holding pattern when airborne.

- **Lateral separation** can be achieved by different means, among which we find:

  - Geographical separation - used in procedural air traffic control, when two aircraft report simultanously that they are overflying different geographic points which are considered by the appropiate ATS authority to be fairly far enough from each other so as to consider the situation to be safe.

  - VOR, NDB or Fixed-based separation - also used in procedural air traffic control by requiring aircraft to fly outbound on specified tracks from the station, which are separated by a minimum angle (15, 30 or 45 degrees depending on whether it is a VOR, NDB or Fixed-based separation). At least one aircraft must be at a distance of 15 nm or more from the facility. See Figure [10.7].

- **Longitudinal separation** refers to the separation between two consecutive aircraft flying on the same track. Depending on the relative ground speeds of

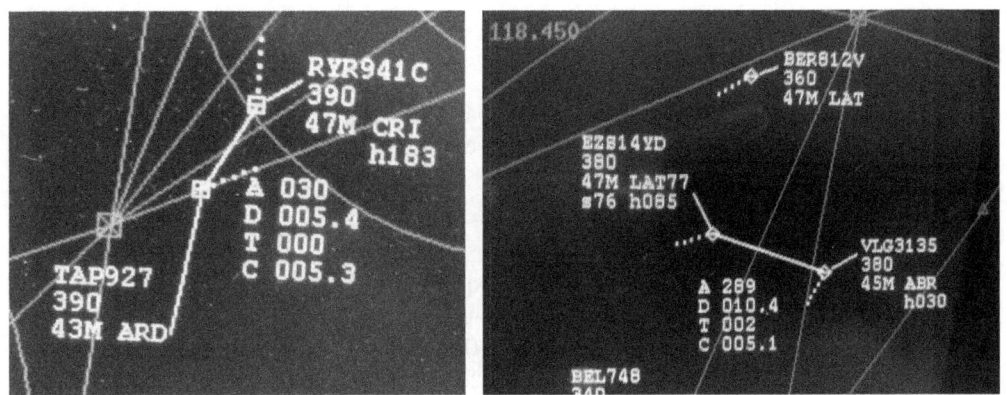

Figure 10.6: RYR941C being vectored left for conflict avoidance with TAP927 (left) and EZS14YD together with VLG3135 being both vectored to the right for conflict resolution (right). Minimum required crossing distance was 5 nm

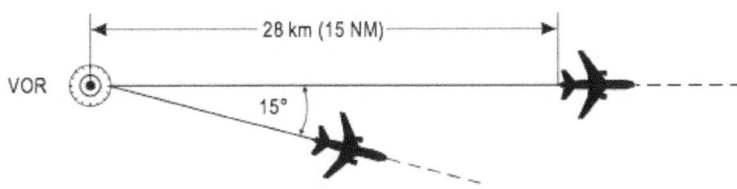

Figure 10.7: VOR-based separation standards © ICAO Doc 4444. *Reproduced with the permission of ICAO*

the aircraft, there might be an opening or closing gap situation. Particular attention must be driven to overtaking situations. Speed control is required to maintain longitudinal separation, typically through the use of the Mach number technique[7]. If it is essential to make an immediate temporary change in the Mach number (e.g. due to turbulence), ATC shall be notified as soon as possible that such a change has been made.

4.1.1 Separation minima based on ATS surveillance systems

ICAO Doc 4444 establishes, in general, a minimum separation of 5 nm between aircraft when advanced surveillance systems are used (such as radar, ADSB or

---

[7]Aircraft are instructed to maintain a minimum, maximum or exact true Mach Number

MLAT). However, national ATS authorities may state different separation minima values, but never below:

- 3 nm for areas where surveillance capabilities permit so (typically in complex TMA environments with high radar accuracy)

- 2.5 nm between two consecutive arrivals when they are less than 10 nm away from the runway threshold

Depending on radar coverage and signal quality, ATS authorities may establish certain areas or periods over which the separation minima changes. This comes determined in the associated AIP. Figure [10.8] shows the different radar separation minima as published in the spanish AIP for different airspace volumes, being a function of radar availability.

3. MÍNIMOS DE SEPARACIÓN RADAR HORIZONTAL.

3. RADAR HORIZONTAL SEPARATION MINIMA.

3.1. Vigilancia radar con primario.

3.1. Primary surveillance radar.

Se establecen los siguientes valores mínimos de separación radar horizontal entre blancos primarios o un primario y un secundario, para un escenario común para los espacios aéreos TMA, APP y TWR:
  PSR-PSR: 5 NM
  PSR-SSR: 5 NM

The following radar horizontal separation minima values are established between primary targets or a primary and a secondary target, for a common scenario for the airspaces TMA, APP and TWR:
  PSR-PSR: 5 NM
  PSR-SSR: 5 NM

3.2. Vigilancia radar en ruta con SSR.

3.2. SSR en-route radar surveillance.

DEPENDENCIA UNIT	TRATAMIENTO DE LA SEÑAL DE VIGILANCIA SURVEILLANCE DATA PROCESSING MODE	SEPARACIÓN SEPARATION
MADRID ACC BARCELONA ACC SEVILLA ACC CANARIAS ACC	Multiradar (remoto y autónomo) Multiradar (remote and autonomous)	5 NM
	Monoradar Civil (autónomo) Civil Monoradar (autonomous)	10 NM
	Monoradar Militar (autónomo) Military Monoradar (autonomous)	15 NM

3.3. Vigilancia radar en APP.

3.3 APP radar surveillance.

DEPENDENCIA (SERVICIO ATS / ESPACIO AÉREO) UNIT (ATS SERVICE / AIRSPACE)	SELECCIÓN RADAR RADAR SETTING		SEPARACIÓN MÍNIMA APLICABLE MINIMUM SEPARATION APPLICABLE		ARP UTILIZADO USED ARP (3)
	Tratamiento Señal Processing Mode	Radar	de/from 0 a/to 30 NM ARP	de/from 30 a/to 60 (1) (2) NM ARP	
BARCELONA ACC (BARCELONA APP / BARCELONA TMA)	Multiradar (remoto y autónomo/ remote & autonomous)	Nominal con / with PSR	3 (4)		LEBL GCLP LEMD LEMG LEPA LEST LEZL GCTS LEVC
GRAN CANARIA ACC (GRAN CANARIA APP / CANARIAS TMA)		Nominal sin / without PSR	3	5	
MADRID ACC (MADRID APP / MADRID TMA)					
MÁLAGA TWR (MALAGA APP / SEVILLA TMA AREA 3)		Degradado / Degraded	5		
PALMA TACC (PALMA APP / PALMA TMA)	Monoradar (autónomo/ autonomous)	Principal / Main Radar	3	5	
SANTIAGO TACC (SANTIAGO APP / GALICIA TMA)					
SEVILLA ACC (SEVILLA APP / SEVILLA TMA)					
TENERIFE SUR TWR (TENERIFE SUR APP / CANARIAS TMA)		No radar principal / Non main radar	5		
VALENCIA TACC (VALENCIA APP / VALENCIA TMA)					

Figure 10.8: Radar separation minima according to ENAIRE AIP ENR 1.6
© ENAIRE

4.1.2 Wake vortex separation

In addition to the already defined separation minima values, which are specified to maintain the integrity of the aircraft and avoid any possible collision, special care must be taken with respect to wake vortex encounters. The fact of entering an area in which the upstream current of air is not still, but instead has been strongly perturbated by a preceding aircraft, may lead to loss of aircraft´s control.

In order to avoid wake vortex encounters, given the fact that these vortexes are invisible, time-based or distance-based separations (depending on the context) are enforced. These separation standards are defined per aircraft pair, taking into account which aircraft generates and which aircraft *suffers* the wake vortex turbulence.

Aircraft are categorized according to their ability to generate and confront wake vortex turbulence into four different groups: Light, Medium, Heavy & Super[8]. Depending on the category of both, the leading and the trailing aircraft, as well as the geometry of the situation, certain minimum separation standards are applied. Table [10.1] summarizes some of the prescribed separation standards for a sequence of departing aircraft from the same runway.

Leading aircraft → Trailing aircraft ↓	Light	Medium	Heavy
Light	0	120	120
Medium	0	0	120
Heavy	0	0	120

Table 10.1: Example of minimum wake vortex separation times (in seconds)

4.1.3 Emergency separation standards

If, during an emergency situation, it is not possible to ensure that the applicable horizontal separation can be maintained, emergency separation of half the applicable vertical separation minimum may be used. This means that a 1000 ft vertical separation minima may be reduced to 500 ft and 2000 ft vertical separation minimum may be reduced to 1000 ft. All flight crews concerned must be advised if emergency separation is used. Additionally, all flight crews concerned shall be provided with essential traffic information.

4.2 Loss of separation

A loss of separation is a violation of the prescribed minima under controlled airspace. An AIRPROX is the code word used in an air traffic incident report

---

[8]Within the Super category we only find the A380

to designate aircraft proximity. A loss of separation may happen in either a vertical or a horizontal plane, or both. There might be losses of separation from an aircraft with respect to other aircraft, the ground or specified volumes of airspace.

The effects of a loss of separation may result in collision. Because of wake turbulence encounters or violent maneuvers executed to avoid aircraft collision, injuries may appear to passengers or crew onboard, who may hit their bodies against some aircraft parts from the inside. Also, the high levels of stress resulting from a loss of separation, both on pilots and air traffic controllers, may result in a reduced working performance.

Eurocontrol defines the following defences to prevent a loss of separation:

- Pilot situational awareness of the location and intent of other aircraft (gained from listening to radio traffic or visual identification) and ACAS, especially when not in receipt of an ATS radar or procedural control or when operating outside controlled airspace

- Standard Operating Procedures, both on the flight deck and in the ATS Unit, which detail procedures to be followed to reduce the risk of loss of separation

- Aircraft onboard equipment which warns of potential collision with other aircraft (ACAS) and allows an appropriate procedural response to risk. However, note that not all aircraft are required to be fitted with ACAS. Only civil turbine-powered aircraft having a maximum certified takeoff mass in excess of 5.700 kg, or a maximum approved passenger seating configuration of more than 19[9]

- Ground-based equipment designed to warn of potential collision with other aircraft (STCA[10])

- Ground-based equipment designed to verify that the current clearances provide adequate separation in the short term (TCT[11])

- Ground-based equipment designed to warn of potential conflict between aircraft in flight (MTCD[12]).

Losses of separation might be controller or pilot-induced.

4.2.1 Conflict definition

A *conflict* is a potential loss of separation which should be detected with enough anticipation, requiring a modification in the relative trajectory of the aircraft to

---

[9]ICAO Annex 6 Part I Chapter 6
[10]Short Term Conflict Alert
[11]Tactical Controller Tool
[12]Medium Term Conflict Detector

be solved. Conflict detection is carried out by the associated ATCOs through the extrapolation of the trajectories of the different pairs of aircraft, foreseeing what will their relative future positions be and whether separation minima infringement is likely to occur.

However, not always do ATCOs succeed in the detection of conflicts. This may happen because of many different causes. ATCOs may experience *blind spots* and detect conflicts only when it is too late or even not detect them at all. A non-detected conflict will evolve into a true loss of separation. A typical case of *blind spot* occurs when an ATCO pays a lot of attention to a pair of aircraft expected to fly close to each other, forgetting about other third parties which may also have a role to play in the situation.

4.2.2 Conflict geometries

Two aircraft at the same altitude or Flight Level may be flying intersecting trajectories with different possible geometries. Figures [10.9] and [10.10] summarize the three possible cases. Each of these cases will be deconflicted differently, as we will see in section (4.3).

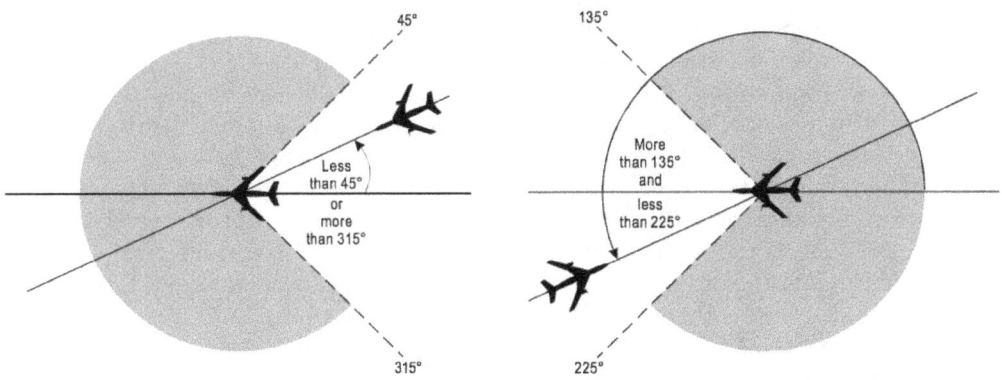

Figure 10.9: Aircraft on same (left) and reciprocal (right) tracks © ICAO Doc 4444. *Reproduced with the permission of ICAO*

4.3 Existing tools and deconflicting procedures

In order to assist the ATCO in the detection of the conflict and in the modification of the aircraft´s trajectory once airborne, different tools are typically integrated in the working position of the controller. The most relevant being the ones described in this section.

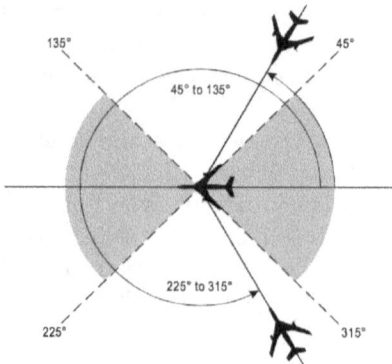

Figure 10.10: Conflict geometry for two aircraft on crossing tracks © ICAO Doc
4444. *Reproduced with the permission of ICAO*

4.3.1 Tactical Controller Tool (TCT)

When integrated together in the ATCO position, the TCT is a software tool that
helps the controller measure the true distance, possible bearings and extrapolated
expected crossing distance between pairs of aircraft. This way, the controller can
easily (with his own mouse) predict what the situation will be like in the short-
term future, according to the current evolution of the different flight variables,
detecting conflicts.

A good Tactical Controller Tool will thus not only calculate the minimum
crossing distance between a pair of aircraft, but will also provide assistance in
the resolution of the conflict, suggesting in which direction and how many degrees
should aircraft turn to solve the conflict. Figure [10.11] shows a TCT indicating
that the distance between the pair of aircraft is 10.4 nm, being the bearing 284
degrees from the VLG3135 to the EZS14YD and expecting a minimum crossing
distance in 2 minutes of 5.1 nm.

4.3.2 Short Term Conflict Alert (STCA)

A STCA is a ground-based safety net intended to assist the controller in preventing
collision between aircraft by generating, in a timely manner, an alert of a potential
or actual infringement of separation minima. Hazardous situations related to aircraft separation can remain unnoticed by the flight crew and the controller. The
controller's workload and priorities may cause an imminent hazardous situation
to remain undetected if not alerted by a STCA. This is especially likely to occur
during heavy workload conditions. For a successful implementation of a STCA, it
is necessary to tune the function and the algorithms taking into account the local
specifications and to provide relevant training to both controllers and engineers.

According to ICAO Doc 4444, the generation of short term conflict alerts must

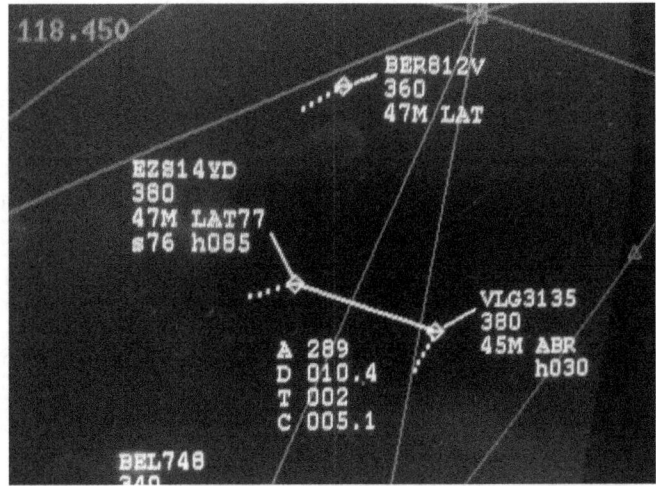

Figure 10.11: Example of a Tactical Controller Tool

be a function of an ATC radar data processing system. The objective of the STCA function is to assist the controller in maintaining separation between controlled flights by generating, in a timely manner (typically 2 minutes before the loss of separation occurs), an alert of a potential infringement of separation minima. In the STCA function the radar-derived current and predicted three-dimensional positions of transponder-equipped aircraft with Mode C capability are monitored for proximity. If the distance between the three-dimensional positions of two aircraft is predicted to be reduced to less than the defined applicable separation minima within a specified time period, an acoustic and/or visual alert will be generated to the radar controller within whose jurisdiction area the aircraft is operating.

In the event an STCA is generated in respect of controlled flights, the controller shall without delay take action to ensure that the applicable separation minima will not be infringed. STCA is a concept rather than a specific system; there are therefore a number of differing implementations used by ATC providers across the world.

Similarly to what happens with TCAS alerts, there can be STCA *false positives*. These *fake alerts* reduce the trend of urgency with which ATCOs react, since they may get used to operating under a framework in which most of the alerts are false positives, acting poorly, hardly paying attention and reducing their confidence in the STCA when a true alert is triggered. Because of this, it may become necessary to inhibit the STCA in some airspace volumes where traffic complexity is so high that the algorithms contained within the STCA do not capture properly the conflict detection criteria or are not properly calibrated.

4.3.3 Medium Term Conflict Detector (MTCD)

MTCD is a flight data processing system designed to warn the controller of potential conflict between flights in his area of responsibility in a time horizon extending up to 20 minutes ahead. MTCD is thus also a ground-based safety net which works similarly to the STCA but with a greater time horizon and taking into account not only the radar-derived information regarding the true trajectory that aircraft are flying, but also the intentions contained within their respective Flight Plans. MTCD is a predictive tool.

MTCD is a natural extension of the Short Term Conflict Alert (STCA) concept. However, one important distinction needs to be made: while STCA is a safety net function and its objective is solely to improve safety, MTCD is also a controller tool which helps conflict resolution and aircraft sequencing with larger time horizons.

4.3.4 Deconflicting procedures

Conflicts are solved forcing modifications of aircraft trajectories. These modifications obbey to clearances issued by ATCOs. Three are the flight parameters that can be modified:

- **Heading** - making use of the so-called *vectoring* technique, defined as the provision of navigational guidance to aircraft in the form of specific headings, based on the use of radar

- **Altitude** or **Flight Level** - in order to achieve vertical separation before horizontal separation is lost

- **Speed** - reducing or increasing the IAS (for levels below FL250) or the Mach Number (above FL250), avoiding overtaking situations

4.4 Letters of Agreement and Tactical Coordination

As we have already explained, each ATCO is responsible for his own piece of airspace (sector) and thus, a transfer of both, control and communication has to be made every few minutes over each aircraft as it passes through the different airspace boundaries existing between the respective sectors or FIRs/UIRs. A relationship and agreement must exist between the transferring and the receiving ATS sectors.

According to ICAO, ATC units should, to the extent possible, establish and apply standardized procedures for the coordination and transfer of control of flights, in order to reduce the need for verbal coordination. Such coordination procedures shall conform to the procedures contained in the following provisions and be specified in letters of agreement and local instructions.

Letters of agreement are documents signed between different ATS Units in which the conditions for the transfer of control are defined (e.g. aircraft leaving LECM UIR via ABRIX into LFBB UIR will be separated longitudinally at least with 10 nm and with their speeds properly adjusted to avoid any overtaking after the transfer of control).

Transfer of communication (change of VHF frequency from one sector to the next one) is typically done before the aircraft reaches the boundary between the two sectors. Transfer of communication must not be mistaken with transfer of control. If transfer of communication is used to transfer an aircraft to a receiving ATC unit, responsibility for control shall not be assumed until the time of crossing of the control area boundary between the sectors. This means that the receiving ATC sector can not *touch* the aircraft (modify its trajectory) until the aircraft is completely under his own jurisdiction, unless a real-time coordination between both ATCOs takes place asking for permission.

5   Phraseologies

Communication between pilots and air traffic controllers is carried out using radio or CPDLC channels. It is necessary to standarize the way in which these communications take place. Phraseologies refer to the way in which the different messages and clearances are transmitted from a linguistic point of view. Communication procedures shall be in accordance with ICAO Volume II of Annex 10 (Aeronautical Telecommunications) and pilots, ATS personnel and other ground personnel shall be thoroughly familiar with the radiotelephony procedures contained therein.

It would not make sense to get too deep into phraseologies within this introductory text. We will just say that messages have to be emitted according to the following structure: *"Name of the station to which the message is intended - Content of the message - Name of the originator of the message"* and that after an instruction is issued to an aircraft, the pilot has to *readback* the clearance so that the ATCO can make sure that the pilot properly received and understood what the ATCO asked him to do. For example:

- *Aeroflot 2500, descend altitude 6000 ft, QNH 1016, Madrid Approach* - message sent by the ATCO instructing the AFL2500 to comply with certain vertical maneuver

- *Madrid, Roger, descending altitude 6000 ft with QNH 1016, Aeroflot 2500* - aircraft reading the message back to the ATCO

Chapter 11. Safety

The Air Transport industry works because users trust aviation as a safe transport mode. If safety were not guaranteed, aviation would not exist as we know it today. We can understand a flight operation as a concatenation of processes that take place one after another (along the different phases of the flight, different actions are taken to overcome potentially threatening situations). Even though many of these actions require human interaction (providing flexibility to the system), most of them are standarized and are part of operational procedures, which are established to keep minimum levels of safety (e.g. cockpit checklist compliance or operational ATM procedures).

Safety is constantly monitored by different stakeholders, mainly regulators (state agencies), airlines and ANSPs. However, when monitoring something (safety in this case), it is necessary to quantify, to measure, to establish certain threshold values that can differentiate a safe situation from an unsafe one. The establishment of these thresholds is not always clear and is normally determined by the degree of danger to the operation encountered as a consequence of an action. Constant monitoring of operations, investigation of incidents and accidents is thus necessary for the determination of these safety thresholds and for the discovery of potential system faults.

Traditional methodologies insist on the research and study of situations where a lack of safety has occured. This forces us to talk about *incidents* and *accidents*. ICAO Annex 13 defines them as follows:

- **Incident** - an occurrence, other than an accident, associated with the operation of an aircraft which affects or could affect the safety of operation

- **Accident** - an occurrence associated with the operation of an aircraft which takes place between the time any person boards the aircraft with the intention of flight until such time as all such persons have disembarked, in which:

  − a person is fatally or seriously injured[1] as a result of:

---

[1] For statistical uniformity only, an injury resulting in death within thirty days of the date of the accident is classified as a fatal injury by ICAO

* being in the aircraft, or
* direct contact with any part of the aircraft, including parts which have become detached from the aircraft, or
* direct exposure to jet blast, except when the injuries are from natural causes, selfinflicted or inflicted by other persons, or when the injuries are to stowaways hiding outside the areas normally available to the passengers and crew

− the aircraft sustains damage or structural failure which:

* adversely affects the structural strength, performance or flight characteristics of the aircraft, and would normally require major repair or replacement of the affected component, except for engine failure or damage, when the damage is limited to the engine, its cowlings or accessories; or for damage limited to propellers, wing tips, antennas, tires, brakes, fairings, small dents or puncture holes in the aircraft skin; or

− the aircraft is missing or is completely inaccessible[2]

1 Reporting systems

The whole air transport industry believes that a lot can be learnt from past safety events, and that their occurence should lead to actions taken for the avoidance of repetition. It is for this reason that ICAO strongly supports the analysis and exchange of incident and accident data.

ICAO Annex 13, through Chapter 8, establishes that all contracting States shall establish mandatory incident reporting systems through which pilots, aircraft mechanics and air traffic control professionals have to report any potential hazard or incident in which they are involved for its analysis. The basic goal of these reporting systems is to increase the size of incident datasets, driving better and more consistent analysis and conclusions. Some professionals may feel afraid that reporting can have punitive consequences against them (they might not want to unveil an incident probably caused because of an unprofessional misconduct or lack of adherence to a certain procedure), however, human errors are sometimes the consequences of other latent causes, and this can only be investigated if incidents are reported and analyzed by research teams. Aware of this problem, institutions are trying to extend the so-called "just culture", that forbids any punitive action against reporting professionals, and tries to protect the sources of information. Figure [11.1] summarizes the way the current safety reporting system is deployed.

ICAO recommended the implementation of Safety Management Systems (SMS) for different Aircraft Operators, Air Navigation Service Providers and supervised

---

[2]An aircraft is considered to be missing when the official search has been terminated and the wreckage has not been located

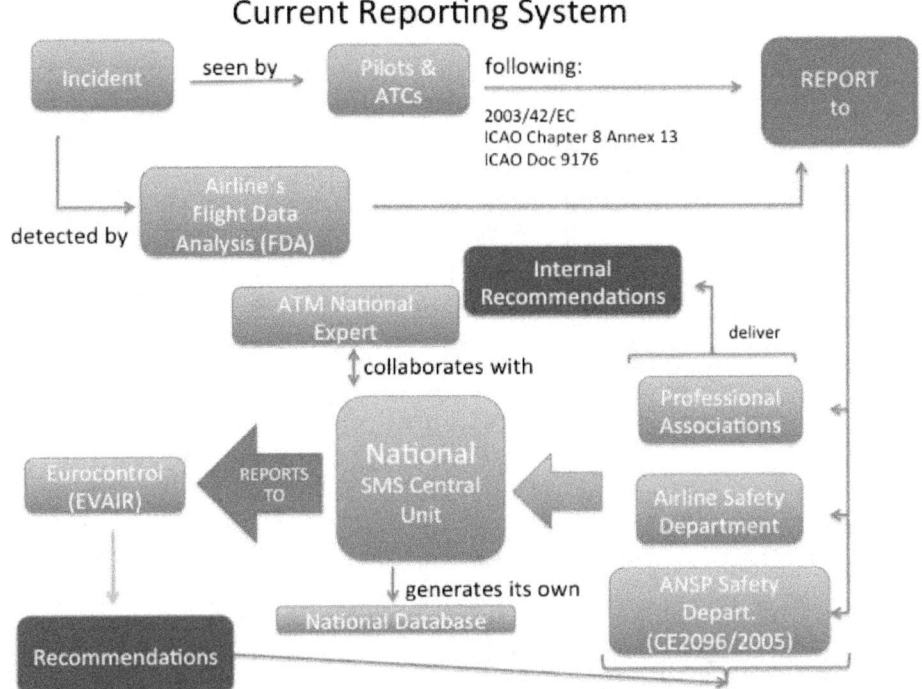

Figure 11.1: Safety reporting system in Europe

by State Nationaly Safety Agencies. The standards under which a SMS has to be implemented are developed in ICAO Annex 19.

2 The Reason Model

The Swiss Cheese model (sometimes referred to as the *cumulative act effect*) is a model that looks for causality relationships behind accidents and incidents, through the use of concepts such as risk analysis and management. This model is used in many different fields, aviation among others (such as healthcare or security).

This model equates each of the different subsystems that conform a system to slices of swiss cheese layered behind each other (see Figure [11.2]). On one side of the slices there are hazards (or potential threats), from where straight-line trajectories are born looking for aligned combinations of cheese holes (representing each of them, different flaws of each subsystem). If opportunity windows are aligned in such a way that let the hazard penetrate all through the swiss cheese, an event finally takes place (accident, incident, or generally, a loss of the system).

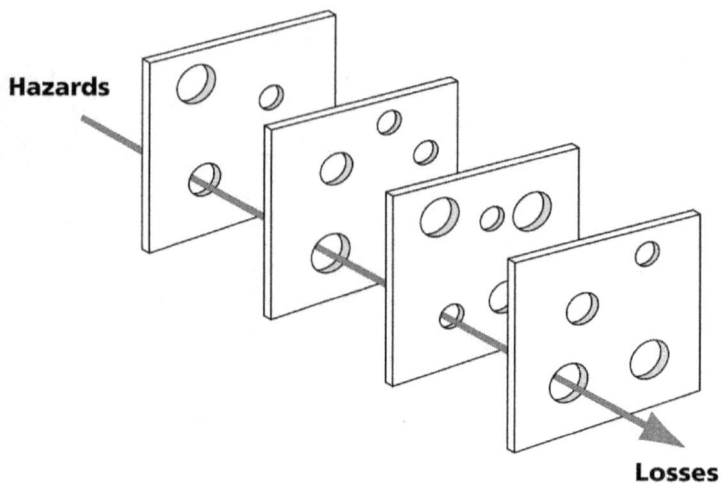

Figure 11.2: Illustration depicting the Reason Model © Davidmack / Wikimedia / CC BY-SA 3.0

As it can be easily understood, the larger and the higher the amount of holes, the less efficient each slice is preserving safety. The model clearly shows how different safety barriers may, when some subsystems fail, prevent the hazard from reaching a loss.

According to Reason, there are two types of failures, **active** or **latent**.

- Active failures - those that have a direct impact on safety, typically human errors. Active failures are specific errors in time and show a clear relationship between cause and consequence.

- Latent failures - they are not so easy to identify, they include factors that may have contributed to a safety event and that may have remained hidden for days, months or even years. Failures in training, culture or procedures are examples of latent failures.

The safety barriers of a system (represented as slices of the swiss cheese) may belong to one or more of the four domains depicted in Table [11.1], according to Reason.

3 Typical ATM-related accidents & incidents

There are many safety aspects within the operation of aircraft from the cockpit perspective (operation of the different aircraft systems). Some of them have a

Domain	Failure Type
Organizational influences	Latent failures
Supervision	
Preconditions	
Specific acts	Active failures

Table 11.1: Subsystem domains and associated failure category

repercusion on ATM since they may require changes in the trajectory of the aircraft (e.g. an emergency descent because of a depressurization). We should then differentiate between events where ATM has primary event causality and events that, despite not having their origin in ATM, have an ATM repercusion. The following list refers to the former, for some cases there is evident relationship between an incident and an accident type, understanding the accident as a severe degradation of its associated incident.

- **Aircraft mid air collision**, happening when two airborne aircraft get in touch. Due to the normal high speed of the vehicles, important structural damage is normally caused, provoking subsequent later impacts (against the sea or ground). Aircraft mid air collisions do, in most cases, have fatal consequences. Examples are the Iberia against Spantax collision over Nantes in 1973[3], the brazilian GOL 737 against a private Embraer over the Amazon in 2006[4] or the Überlingen accident in 2002[5].

- **Loss of separation between a pair of aircraft (AIRPROX)** is, according to ICAO Doc 4444, *a situation in which, in the opinion of a pilot or air traffic services personnel, the distance between aircraft as well as their relative positions and speed have been such that the safety of the aircraft involved may have been compromised.* If the situation had degraded, it could have led to an aircraft mid air collision. Depending on the severity of the loss of separation, these events have different categories (see Table [11.2]). Methodologies have been developed in incident investigation to classify AIRPROX events.

- **Aircraft against object collision on the ground**, which may happen at different parts of the airport, being more critical the faster the aircraft is moving. Runway collisions at high speed may end up with aircraft destruction, while collisions on the apron or taxiways may not be so critical, despite

---

[3]The final report by the French authorities can be found in http://www.bea.aero/docspa/1973/ec-c730305/pdf/ec-c730305.pdf

[4]Final report to be found in http://www.skybrary.aero/bookshelf/books/546.pdf

[5]Report can be found in http://www.skybrary.aero/bookshelf/books/414.pdf

Category	Definition
A	Risk of collision
B	Safety not assured
C	No risk of collision
D	Risk not determined

Table 11.2: Airprox categorization

of creating important structural damage to the aircraft. Objects against which an aircraft can crash are typically other aircraft[6] or vehicles[7].

- **Runway incursion**[8] is defined by ICAO as *any occurrence at an aerodrome involving the incorrect presence of an aircraft, vehicle or person on the protected area of a surface designated for the landing and take off of aircraft*, occuring typically as a pilot error to comply with an ATC clearance or as an ATC error when instructing an aircraft or a vehicle to enter a runway that is already assigned to a different landing or take off operation. Runway incursions may lead to ground collisions, rejected take off maneuvers or go arounds. Runway incursions are typically dangerous when low visibility prevents both pilots and ATCOs to realize the presence of obstacles on the runway. It must be clearly stated that safety barriers such as TCAS do not solve runway incursion situations. Eurocontrol defines the following scenarios as the most common under which runway incursions occur.

  - Departing aircraft runway entry contrary to ATC clearance

  - Aircraft runway crossing after landing contrary to ATC clearance

  - Issued ATC taxi clearance in conflict with another ATC clearance

  - Towed aircraft runway crossing contrary to ATC clearance

Given the high level of potential risk of collision because of runway incursions, ICAO has developed a Manual for the Prevention of Runway Incursions (ICAO Doc 9870).

- **Runway excursion** is defined, according to ICAO, as an overrun off the runway surface. It happens during take off or landing maneuvers, when required runway length does not match available distances. For departures, most common scenarios are aircraft unable to reach rotation before the end

---

[6]The worst accident ever in aviation was a runway collision during take off of two Boeing 747s in Tenerife Norte airport, the final report can be found in http://www.skybrary.aero/bookshelf/books/809.pdf

[7]Recently, a Falcon 50 with French petrol company Total CEO onboard crashed against a snow machine while at take off in Moscow, last October 20th 2014

[8]An example of a runway incursion can be found in youtube.com/watch?v=1N5THRSp4hM

of the runway or rejected take off maneuvers that overrun the end of runway while trying to stop the aircraft. For landings, inability to brake the aircraft before the end of the runway (plus other contributing factors such as for instance, a late touchdown on the runway). Although most factors contributing to a runway excursion have to do with the performance of the aircraft and pilot decision making (e.g. weight of the aircraft, reduced thrust, performance computations, slope of the runway, etc), ATM may play a key role contributing to runway excursions, particularly because of a wrong runway choice (a decision that is made by the tower air traffic controller) with significant tailwind component (that can contribute to aircraft landing faster with respect to the ground), or because of a late or absent transmission of relevant information (windshear, runway braking action condition, etc).

- **Controlled flight into terrain (CFIT)**[9] occurs when an airworthy aircraft (that has no control or structural problem, and is not stalled), under the complete control of the pilot, is inadvertently flown into terrain or water. There is a lack of situational awareness (specially in the vertical plane) until it is too late. Most CFIT accidents are associated to non-precision approaches (where no vertical guidance is provided by any navaid, and pilots have to follow procedures depicted on charts, based on vertical rates, distances, etc) under IMC conditions or during the night. The most important defense against CFIT is a good situational awareness, that can be brought back to the pilot thanks to the GPWS (Ground Proximity Warning System), Standard Operational Procedures (adherence to procedures, checklist, etc) or ATC (advising aircraft tresspassing minimum defined altitudes for terrain separation).

- **Loss of minimum required separation between aircraft and terrain** are safety events, where if situation awareness is not recovered and inmediate action is taken, can lead to a CFIT. They can be pilot or ATC induced, normally because of wrong altimeter setting, level bust, incorrectly assigned altitude (below minimum safe altitudes) or incorrect communication readback (pilot hearing a cleared altitude that is not the one intended by ATC). Minimum altitudes come typically determined by:

  - MSA (Minimum Sector Altitude) - with a radius of 25 nm from the navaid with respect to which a certain approach procedure is designed, for different sectors (determined by radials), minimum altitudes are defined to avoid terrain interference.

  - MVA (Minimum Vector Altitude) - altitudes defined typically in a TMA below which approach air traffic controllers must not vector aircraft, since separation with terrain is not guaranteed.

---

[9]A CFIT example is the Avianca B747 accident during the approach to Madrid-Barajas in November 1983

– MEA (Minimum Enroute Altitude) - values given for each ATS route to assure (among other factors) obstacle clearance.

- Very low **fuel level (emergency)** can be the consequence of airspace saturation or wrong flight planning. The amount of fuel to be loaded onboard comes determined by law and emergencies are declared when the expected amount of fuel onboard when at landing provides less than 30 minutes of flight. Recent events[10] have raised concerns about fuel planning. Since extra fuel carriage costs money to the airlines (cost to carry), they have developed policies to load as less fuel as necessary. Unexpected events (such as a runway blockage, storms stronger than expected or others) can lead to instruct aircraft to hold over their destination aerodromes, wasting fuel and requiring extra flight time. Before declaring a fuel emergency, a deviation to an alternate aerodrome is normally requested, or priority because of low fuel level.

- **Penetration of aircraft into non-authorized airspace** happens when there is a mismatch between aircraft trajectory and ATC clearance. This increases the uncertainty of the ATM system and particularly affects air traffic controllers, who normally do not know the reason (if not communicated) of the deviation. These deviations may cause aircraft to enter D, P or R areas (that might be activated and with military activity) or AIRPROX events (if coming too close to other aircraft). Vertical or lateral deviations from flight plan or ATC instructions may lead to this kind of safety events. A common non-authorized penetration is a VFR traffic with a non-professional pilot flying within unfamiliar airspace (without transponder or radio communication). These airspace penetrations are particularly dangerous in the vicinity of big airports, where IFR traffic is intense.

- **Deviation from lateral ATC clearance** can have similar consequences to the penetration of non-authorized airspace, that is, getting closer to other aircraft (AIRPROX) or ending up penetrating restricted areas. These deviations are normally caused because of wrong readbacks or wrong FMS flight planning (the flight plan that the crew has loaded onboard the FMS does not match the flight plan the ATC has, reflecting the future intentions of the aircraft).

- **Level Bust** is a deviation from vertical ATC clearance. As we saw in our altimetry lesson, level bust incidents can lead to losses of vertical separation between aircraft (AIRPROX) and may trigger TCAS. Level Busts happen when aircraft surpass their cleared flight level while climbing or descending. Aircraft momentum may increase the risk of a level bust, being thus advisable to reduce vertical speed when approaching a cleared altitude or flight level.

---

[10]For more information one may read http://avherald.com/h?article=454af355

Typical causes behind level busts are wrong readback, wrong selected altitude or wrong altimeter setting.

- **Non-stabilized approaches**[11] are those in which the energy of the aircraft is higher than recommended (aircraft flying too high or too fast in the last phases of the approach) and may lead to a missed approach or runway excursion. ATC do normally impact these events when forcing pilots to make maneuvers that are incompatible with a smooth and quiet aircraft performance, particularly when shortcutting their trajectories during the approach, forcing them to descend many levels in a short horizontal distance (accelerating the aircraft). The establishment of target points with their respective speeds and altitudes mitigates risks for a non-stabilized approach. Pressure and stress set on the crew to fit into a sequence of arrivals may lead to an excess of aircraft speed.

- **Engine fire/shutdown** is specially critical during take off, requiring the aircraft to fly back to the airfield. The repercusion from the ATM perspective is huge since full priority has to be given to the aircraft (sometimes landing over its certified Maximum Landing Weight and maybe blocking the runway). A complete change in the sequence of use of the runway has to be made.

- **Hydraulic emergencies** may lead to aircraft with turning problems (they might only be able to turn in one direction), landing gear extension problems and possibly, runway blockage.

- **Bird impacts** are very common in the vicinity of aerodromes and the reason for which airport falconry makes sense. Large birds (e.g. eagles or vultures) ingested by aircraft engines may lead to engine malfunctioning or shutdown, aircraft declaring emergency. Other impacts can provoke structural damage and leave debris on the runway, that must be inspected after a bird impact is reported.

Among the most common **causes** behind these incidents, we may find failures in the following domains:

- **Wrong ATC readback** - when ATC instructions are transmitted to pilots (via VHF radio communication), pilots have to read them back to ATC to ensure they have properly received and understood what they are instructed to do. Sometimes, pilots understand and readback an instruction different from the one that is given by the ATC, not being this former one aware of the wrong readback. Aircraft act thus according not to what they have been instructed to do, but to what they have said they will do according to their readback. A couple of examples of communications in which a wrong readback occurs are now described:

---

[11]An example can be found in https://www.youtube.com/watch?v=EwSSWg3K9Qw

- ATC: *RYR4WH, descend FL210*
- RYR4WH: *Copied, descending FL200* -> generating a level bust as a result

- ...

- ATC: *KLM13T, due to traffic, descend now FL390, traffic at your level*
- EZY43T: *Copied, descending FL390* -> wrong station reads back because of similar callsigns.

- **Fatigue** - both in pilots and air traffic controllers, fatigue reduces their ability to make the right decisions, acting more based on natural instincts than on good reasoning. Pilot disorientation and loss of situational awareness are strongly related to fatigue. Because of this, mandatory resting times are enforced to pilots and air traffic controllers, to preserve their minds as fresh as possible.

- **Poor CRM coordination** - Crew resource management has been proved crucial in aviation safety. All cockpit pilots must act as a team when solving situations and making decisions. An authoritative captain attitude led to important accidents and safety incidents in the 80s. This is still today a major concern in some asian growing economies.

- **Lack of adherence to ATC instructions**, lateral or vertical, can lead to AIRPROX. Specially concerning events in which pilots try to assure separation with other aircraft visually, third parties may interact (taken into account by the ATC but not by the pilot) and provoke a loss of separation with this third aircraft involved.

- **Deficient coordination between different ATS units** - when non standard clearances or maneuvers have to be put in place (e.g. because of presence or CBs), constant coordination is required between different sectors, concerning the conditions under which traffic will be transferred by the different air traffic controllers. On some occassions, it does not end up clear which is the solution strategy to solve a conflict, specially when problems arise at sector borderlines (air traffic controllers may not agree on who is going to do what).

- **Airspace distribution poorly designed** is a very critical latent failure, which may remain hidden for months. Airspace poorly designed can lead to aircraft entering sectors without proper surveillance or situational awareness of air traffic controllers or other pilots.

- **Human errors** are a constant within the aviation system. Any time we put human beings into the equation (from design, to construction, maintenance and tactical operation), we may be introducing errors associated to their decisions and designs. However, the important thing should not be to eliminate the human being from the chain (which, well understood, is totally impossible since planes and systems will always be designed by humans), but to detect these errors in time and put the necessary means to avoid the consequences. From a pilot and air traffic controller perspective, many human errors are induced by the "tunnel vision" or the "blind spot fact", which basically consists of a human agent focusing all his efforts on solving a determined problem, not realizing that another third agent is playing a key role in the problem scenario, and that has to be taken into account when developing a solving strategy. When third parties are involved and not taken into account (e.g. a third converging aircraft when solving a conflict between a couple of aircraft), there is a risk that the chosen solution may create a new conflict against this third agent.

- **Lack of knowledge** together with deficient training programs may lead professionals to act following their own instincts instead of their flight or control education principles, making things worse and not solving the problems that may arise.

- **Lack of adherence to procedures**, excessive pilot or air traffic controller confidence may lead to wilful misconducts.

- **Poorly designed procedures** that may lead to contradictory orders or may create conflicts instead of solving them. They belong to the latent failure domain and are thus, of critical importance. Such cases are reflected on texts such as Operation Manuals, Letters of Agreement or AIP Procedures.

Chapter 12. Cartography

From the beginning of aviation, common needs with maritime navigation required pilots to identify and know their position in territories with which they were not familiar. Initially, maps (for visual navigation) and later navigation charts (including more detailed information such as navaids, minimum altitudes or flight procedures) became totally necessary to understand designed trajectories and routes. Currently, despite the extended use of digital presentations, the use of navigation charts is still required. The introduction of the Electronic Flight Bag, that aims at reducing paper weight inside the cockpit, represents current state of the art in the use of air navigation cartography. Understanding geographic tracks and distances is basic for flight planning, playing navigation charts and tools a key role. A basic knowledge is required to understand aeronautical charts, avoiding misunderstandings due to angle or distance distortions.

1 Principles of cartography

The morphological shape of the Earth is very complex and difficult to describe. We use coordinates and their underlying Coordinate Reference System (CRS) to describe the geographic position of an object. We will talk about latitude (angle in degrees with respect to the equator, φ, can be North or South) and longitude (angle in degrees with respect to the Greenwich meridian, λ, can be East or West). To model the Earth, we use ellipsoids (and the sphere), whereas geoids[1] or other gravity related models are, in other applications, the main reference system for elevation (h)[2]. We may use global or local ellipsoids to model the Earth, depending on whether we want to approach the whole shape of the Earth or just the curvature of a certain area. The standard adopted model is the WGS84, which is the world geodetic system we are used to working with in our daily life. It comes determined by its origin, scale, orientation and principal planes.

---

[1] A geoid is an equipotential surface that would coincide with the mean ocean surface of the Earth if the oceans and the atmosphere were in equilibrium, at rest with respect to the rotating Earth

[2] As we have already seen in Chapter 3, in air navigation we determine vertical positions through barometric principles

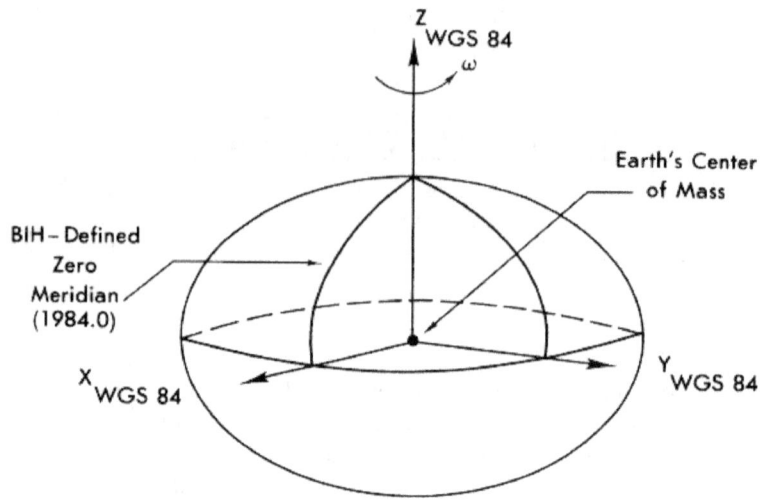

Figure 12.1: WGS84 model of the Earth © Public Domain / Wikimedia

2   Mapping surfaces and projections

We use map projections to represent the surface of a mathematical Earth model (defined ellipsoid) onto a plane based on geometrical or mathematical rules, principles or constraints. Mapping is a process of representation that lets us change from geodetic (φ, λ) to projected coordinates (x, y). This process has its associated pros and cons:

- Advantages - calculation of distances, angles, directions and areas may be made based on the classical geometry rules (Euclidean geometry)

- Drawbacks - geometric distortions appear. They will depend on the point and method of projection. It is not possible to represent a curved surface, like an ellipsoid or a sphere, onto a plane without distortions.

The projection process requires the use of a mapping surface[3], against which and from a projection point, the Earth´s model is projected. Different projections will be obtained depending on the type of mapping surface (plane, cone or cylinder), the coincidence between the mapping surface and the ellipsoid (tangent or secant), and the alignment of the mapping surface with respect to the principal planes of the ellipsoid (normal, transversal or oblique). Also, the point from which the projection is executed characterizes the projection, being gnomonic (projection point is the center of the ellipsoid), stereographic (projection from a point on the surface of the ellipsoid) or orthographic (from a point placed infinitely far away).

---

[3]Typically a cone, a cylinder or a plane, necessarily a developable surface (flattenable)

Distortions are minimal in the tangent or secant points between the ellipsoid and the mapping surface. The following features are defined regarding the behaviour of the projection in terms of distortions:

- Conformity - no angular distortion appears

- Equivalency - areas are preseved

- Equidistancy - specific lines, such as meridians, are mapped with true length

2.1 Lambert Conformal Conic transformation

This projection is obtained by resting (tangent) or intersecting (secant) a cone over or with the ellipsoid that models the Earth. In the LCC projection the tangent parallel between the ellipsoid and the cone is assigned unitary scale and is called the reference or standard parallel. If the cone is secant (it cuts the ellipsoid twice), both of them can be assigned a unitary scale, with this scale decreasing between the two parallels and increasing outside of them.

This projection (see Figure [12.2]) is often used in aeronautical charts, as a straight line drawn on it approximates a great-circle route (orthodromic) between two points (as long as distances are not great). It is also useful to measure angles, since it is a conformal projection. Distortion in terms of distances is not excessive if used for areas within two reference parallels that are not chosen too far from each other. That is, for areas of small latitudinal width, scale is nearly constant.

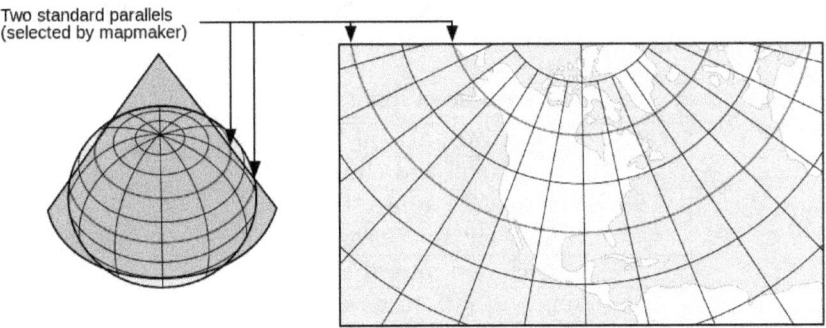

Figure 12.2: Lambert Conformal Conic projection © Public Domain / Wikimedia

2.2 Universal Transverse Mercator (UTM)

The UTM projection is based on the simple Transverse Mercator, a cylindrical projection in which the cylinder is placed parallel to the equator (transverse) and tangent to a meridian. The simple Transverse Mercator projection suffers North-South minimal distortion along the tangent meridian, but distortion increases rapidly with longitude (measured from the tangent meridian). Angles between meridians and parallels are highly distorted as well. It might be adequate to represent areas of elongated shape in the direction of the meridian.

The Universal Transverse Mercator is not a single map projection; it is a 2D Earth grid coordinate system based on the division of the Earth into 60 zones, each a 6-degree band of longitude, and uses a secant transverse Mercator projection in each zone. By properly choosing the central meridian, highly accurate local maps (of narrow width) can be made. North-South distortion is kept to a minimum for each of these zones, and East-West distortion inherent to ordinary Transverse Mercator projection is diminished by defining many narrow sectors and overlapping them. Over ±80 degrees of latitude, it is complemented with the stereographic projection.

These projections represent loxodromes (constant geographic track trajectories) as straight lines, since they are conformal. Distortions in terms of lengths and sizes are minimized through the use of different cylinders. An example of this projection is shown on Figure [12.3].

3   Aeronautical Charts

Due to the importance of a right understanding of navigation charts, ICAO Annex 4 establishes standards and recommended practices regarding Aeronautical Charts. Due to the different nature of navigation along the different flight phases (see Table [12.1]), requirements vary, generating different chart types.

Phase of flight
Taxi from stand to take-off holding point
Take-off and climb
En-route ATS route structure
Descent to approach
Approach and missed approach
Landing and taxi to aircraft stand

Table 12.1: Flight phases according to ICAO Annex 4, for each of which different charts are required

Each type of chart shall provide information appropiate to the phase of flight, ensuring a safe and expeditious operation of the aircraft. The presentation shall be accurate, free from distortion, unambiguous and readable under normal operating conditions (natural and artifical light). Information presented on the charts shall be easily read by the pilots, that in an amount of time compatible with workload, should understand the information that is presented. The information presented

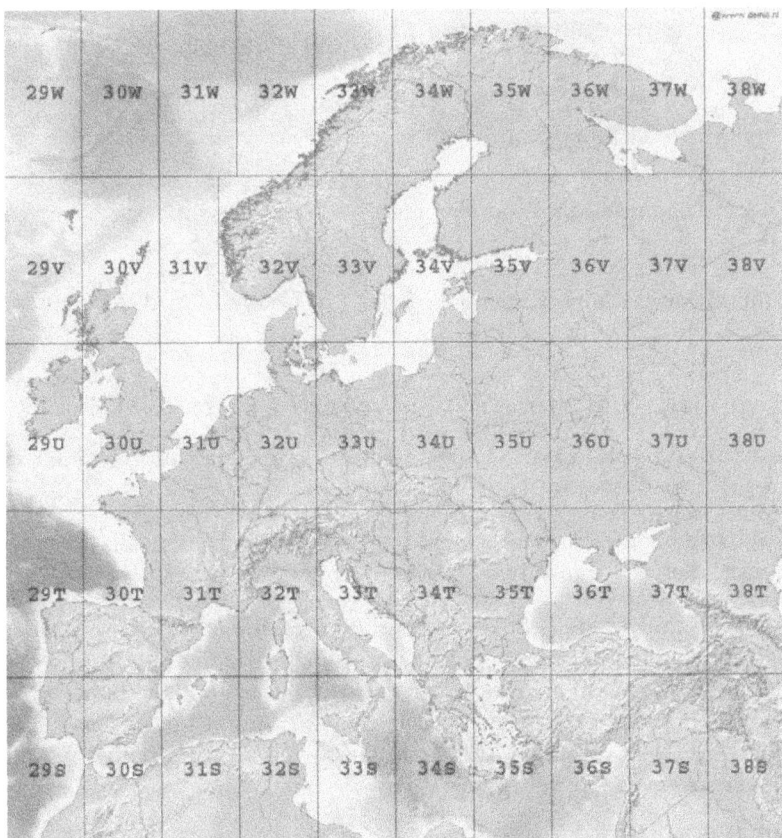

Figure 12.3: Different UTM areas over Europe © Public Domain / Wikimedia

shall make it easy to transition from one type of chart to the next one according to the evolution of the different flight phases. ICAO recommends all charts to be True North orientated (this way, charts will be more constant in time, varying only the depicted local magnetic variation).

Aeronautical Charts will not only represent different territories, but also include relevant information for navigation, including ATS Routes, navaids, ATS frequencies, airspace classes, navigation procedures, presence of danger or restricted areas, etc.

Data consistency is of very high relevance with respect to the safety of the flight. All agents involved in the operation must have the last updated procedures (according to the last published AIRAC cyle, and NOTAMs). This affects Aeronatical Charts, that have to be always updated to avoid confusion and lack of adherence.

ICAO establishes a set of official mandatory charts that must be published by

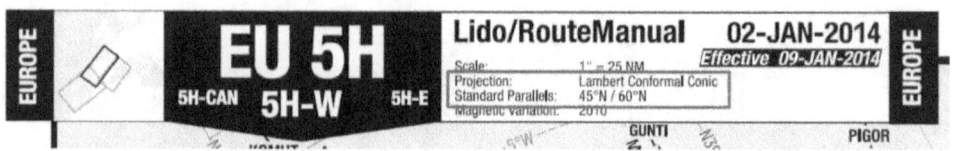

Figure 12.4: Example of a Lido Navigation Chart showing type of projection

each contracting State, and whose guidelines can be found in ICAO Annex 4. We will here focus on the most relevant.

3.1 Standard Instrument Departure Chart (SID)

- **Function** - the SID Chart shall provide the crew with information to comply with the designated Standard Departure route.

- **Availability** - the Standard Departure Chart shall be available wherever a standard departure route (IFR) has been established.

- **Coverage and scale** - the coverage shall be sufficient to show both, the SID initial (normally at the end of the runway) and SID final point (of connection with the assigned ATS route for the enroute phase of the flight). The chart should be drawn to scale, showing a scale-bar, and if not drawn to scale, the annotation "NOT TO SCALE" shall be shown, together with the symbol for scale-break where appropiate.

- **Projection** - conformal projections on which straight lines approximate great circle (orthodromic) trajectories should be used. If chart is drawn to scale, parallels and meridians should be shown at suitable intervals.

- **Identification** - the chart must be identified by the name of the city or town to which the aerodrome serves, and when appropiate, the runway designator and the designator of the standard departure route. When a SID is designed for RNAV, the additional abbreviation "RNAV" shall be given.

- **Culture and topography** - If the chart is drawn to scale, shore lines of all open water areas, lakes and rivers shall be shown unless they conflict with data more applicable to the function of the chart. ICAO recommends to show all relief exceeding 1000ft above the aerodrome elevation through smoothed contour lines, and where appropiate, spot elevations including any significant obstacle, for the sake of a better pilot situational awareness.

- **Magnetic variation** used in determining magnetic bearings, tracks and radials shall be shown to the nearest degree.

- **Bearings, tracks and radials** shall be magnetic, except for areas of very high latitude, where it is determined by the appropiate authority that reference to Magnetic North is impractical. If any bearing, track or radial is given with respect to the True North, it must be clearly specified.

- **Aeronautical data**

  - Aerodrome of departure shall be shown through the use of the runway pattern. All aerodromes that affect the SID shall be shown and identified.

  - Prohibited, restricted and danger areas which may affect the execution of the procedures shall be shown with their identification and vertical limits.

  - Minimum sector altitudes shall be clearly shown, with a clear indication of the sector to which it applies. If they have not been determined, the chart shall be drawn to scale and area minimum altitudes shall be shown within quadrilaterals formed by parallels and meridians.

  - Route designator, significant points of the route, track and distance of each segment, minimum altitude restrictions, significant navaids (including identification, frequency and geographical coordinates), applicable holding patterns, transition altitude, speed restrictions.

  - A textual description of the SID is recommended.

An example of a SID Chart and associated text can be found in the next two pages, published by Enaire (Spanish ANSP) for the airport of Valencia (LEVC).

CARTA DE SALIDA NORMALIZADA
VUELOS POR INSTRUMENTOS (SID)-OACI

VALENCIA
RWY 30

TA 6000

APP 120.100
TWR 118.550

ALT4A   ASTRO4F   CENTA5A
MANDY5A   NARGO2A   ODSEN1A
ODSEN1H   ORVUS1A   ORVUS1H
SOPET4A   TATOS5A

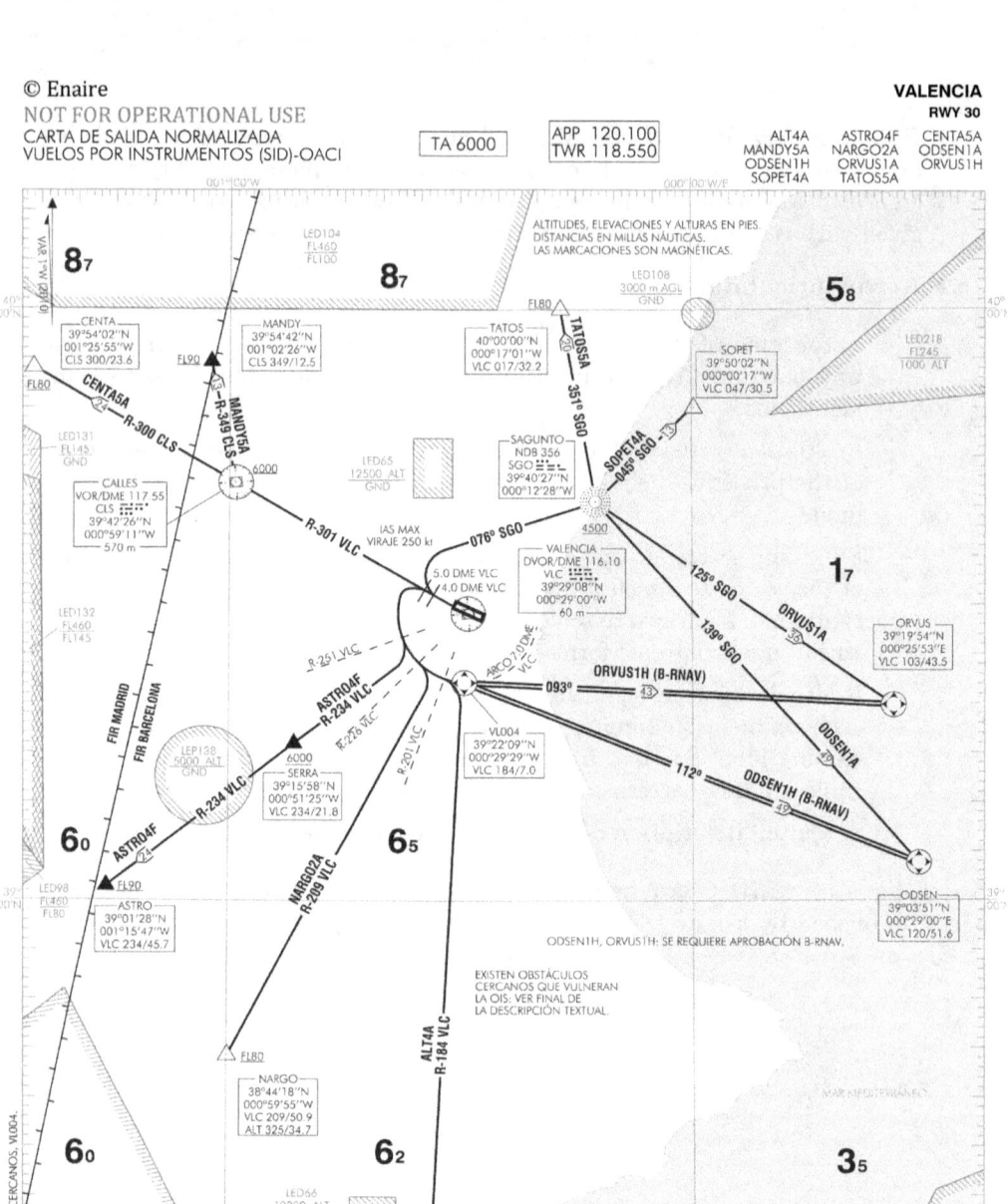

ALTITUDES, ELEVACIONES Y ALTURAS EN PIES.
DISTANCIAS EN MILLAS NÁUTICAS.
LAS MARCACIONES SON MAGNÉTICAS.

ODSEN1H, ORVUS1H: SE REQUIERE APROBACIÓN B-RNAV.

EXISTEN OBSTÁCULOS
CERCANOS QUE VULNERAN
LA OIS: VER FINAL DE
LA DESCRIPCIÓN TEXTUAL.

AIP
ESPAÑA

© Enaire
NOT FOR OPERATIONAL USE

AD 2 - LEVC SID 2.3
WEF 03-MAY-12

VALENCIA AD

SALIDAS NORMALIZADAS POR INSTRUMENTOS (SID)	STANDARD INSTRUMENT DEPARTURES (SID)

PISTA 30 / **RUNWAY 30**

SALIDA ALICANTE CUATRO ALPHA (ALT4A) / ALICANTE FOUR ALPHA DEPARTURE (ALT4A)

Subir en rumbo de pista hasta 5,0 DME VLC. Virar a la izquierda para seguir arco 7,0 DME VLC hasta cruzar R-201 VLC. Virar a la derecha para interceptar y seguir R-184 VLC directo a VOR/DME ALT.
Pendiente mínima de ascenso 5,5% hasta 2000 ft.

Climb on runway heading to 5.0 DME VLC. Turn left to follow arc 7.0 DME VLC to cross R-201 VLC. Turn right to intercept and follow R-184 VLC direct to VOR/DME ALT.
5.5% minimum climb gradient to 2000 ft.

SALIDA ASTRO CUATRO FOXTROT (ASTRO4F) / ASTRO FOUR FOXTROT DEPARTURE (ASTRO4F)

Subir en rumbo de pista hasta 5,0 DME VLC. Virar a la izquierda para seguir arco 7,0 DME VLC hasta cruzar R-251 VLC. Virar a la derecha para interceptar y seguir R-234 VLC directo a cruzar SERRA a 6000 ft o superior. Directo a cruzar ASTRO a FL90 o superior.
Pendiente mínima de ascenso 5,5% hasta 6000 ft.

Climb on runway heading to 5.0 DME VLC. Turn left to follow arc 7.0 DME VLC to cross R-251 VLC. Turn right to intercept and follow R-234 VLC direct to cross SERRA at 6000 ft or above. Direct to cross ASTRO at FL90 or above.
5.5% minimum climb gradient to 6000 ft.

SALIDA CENTA CINCO ALPHA (CENTA5A) / CENTA FIVE ALPHA DEPARTURE (CENTA5A)

Subir en rumbo de pista hasta 5,0 DME VLC. Virar a la derecha para interceptar y seguir R-301 VLC directo a cruzar VOR/DME CLS a 6000 ft o superior. Proceder por R-300 CLS directo a cruzar CENTA a FL80 o superior.
Pendiente mínima de ascenso 5,5% hasta 6000 ft.

Climb on runway heading to 5.0 DME VLC. Turn right to intercept and follow R-301 VLC direct to cross VOR/DME CLS at 6000 ft or above. Proceed on R-300 CLS direct to cross CENTA at FL80 or above.
5.5% minimum climb gradient to 6000 ft.

SALIDA MANDY CINCO ALPHA (MANDY5A) / MANDY FIVE ALPHA DEPARTURE (MANDY5A)

Subir en rumbo de pista hasta 5,0 DME VLC. Virar a la derecha para interceptar y seguir R-301 VLC directo a cruzar VOR/DME CLS a 6000 ft o superior. Proceder por R-349 CLS directo a cruzar MANDY a FL90 o superior.
Pendiente mínima de ascenso 5,5% hasta FL90.

Climb on runway heading to 5.0 DME VLC. Turn right to intercept and follow R-301 VLC direct to cross VOR/DME CLS at 6000 ft or above. Proceed on R-349 CLS direct to cross MANDY at FL90 or above.
5.5% minimum climb gradient to FL90.

SALIDA NARGO DOS ALPHA (NARGO2A) / NARGO TWO ALPHA DEPARTURE (NARGO2A)

Subir en rumbo de pista hasta 5,0 DME VLC. Virar a la izquierda para seguir arco 7,0 DME VLC hasta cruzar R-226 VLC. Virar a la derecha para interceptar y seguir R-209 VLC directo a cruzar NARGO a FL80 o superior.
Pendiente mínima de ascenso 5,5% hasta 2000 ft.

Climb on runway heading to 5.0 DME VLC. Turn left to follow arc 7.0 DME VLC to cross R-226 VLC. Turn right to intercept and follow R-209 VLC direct to cross NARGO at FL80 or above.
5.5% minimum climb gradient to 2000 ft.

SALIDA ODSEN UNO ALPHA (ODSEN1A) / ODSEN ONE ALPHA DEPARTURE (ODSEN1A)

Subir en rumbo de pista hasta 4,0 DME VLC. Virar a la derecha (IAS MAX en viraje 250 kt) para interceptar y seguir ruta magnética 076° SGO directo a cruzar NDB SGO a 4500 ft o superior. Proceder por ruta magnética 139° SGO directo a ODSEN.
Pendiente mínima de ascenso 5,5% hasta 2000 ft.

Climb on runway heading to 4.0 DME VLC. Turn right (turning IAS MAX 250 kt) to intercept and follow magnetic track 076° SGO direct to cross NDB SGO at 4500 ft or above. Proceed on magnetic track 139° SGO direct to ODSEN.
5.5% minimum climb gradient to 2000 ft.

SALIDA ODSEN UNO HOTEL (ODSEN1H) B-RNAV. Se requiere aprobación B-RNAV / ODSEN ONE HOTEL DEPARTURE (ODSEN1H) B-RNAV. B-RNAV approval required

Subir en rumbo de pista hasta 5,0 DME VLC. Virar a la izquierda para seguir arco 7,0 DME VLC directo a VL004. Directo a ODSEN.
Pendiente mínima de ascenso 5,5% hasta 2000 ft.

Climb on runway heading to 5.0 DME VLC. Turn left to follow arc 7.0 DME VLC direct to VL004. Direct to ODSEN.
5.5% minimum climb gradient to 2000 ft.

SALIDA ORVUS UNO ALPHA (ORVUS1A) / ORVUS ONE ALPHA DEPARTURE (ORVUS 1A)

Subir en rumbo de pista hasta 4,0 DME VLC. Virar a la derecha (IAS MAX en viraje 250 kt) para interceptar y seguir ruta magnética 076° SGO directo a cruzar NDB SGO a 4500 ft o superior. Proceder por ruta magnética 125° SGO directo a ORVUS.
Pendiente mínima de ascenso 5,5% hasta 2000 ft.

Climb on runway heading to 4.0 DME VLC. Turn right (turning IAS MAX 250 kt) to intercept and follow magnetic track 076° SGO direct to cross NDB SGO at 4500 ft or above. Proceed on magnetic track 125° SGO direct to ORVUS.
5.5% minimum climb gradient to 2000 ft.

SALIDA ORVUS UNO HOTEL (ORVUS1H) B-RNAV. Se requiere aprobación B-RNAV / ORVUS ONE HOTEL DEPARTURE (ORVUS1H) B-RNAV. B-RNAV approval required

Subir en rumbo de pista hasta 5,0 DME VLC. Virar a la izquierda para seguir arco 7,0 DME VLC directo a VL004. Directo a ORVUS.
Pendiente mínima de ascenso 5,5% hasta 2000 ft.

Climb on runway heading to 5.0 DME VLC. Turn left to follow arc 7.0 DME VLC direct to VL004. Direct to ORVUS.
5.5% minimum climb gradient to 2000 ft.

SALIDA SOPET CUATRO ALPHA (SOPET4A) / SOPET FOUR ALPHA DEPARTURE (SOPET4A)

Subir en rumbo de pista hasta 4,0 DME VLC. Virar a la derecha (IAS MAX en viraje 250 kt) para interceptar y seguir ruta magnética 076° SGO directo a cruzar NDB SGO a 4500 ft o superior. Proceder por ruta magnética 045° SGO directo a SOPET.
Pendiente mínima de ascenso 5,5% hasta 2000 ft.

Climb on runway heading to 4.0 DME VLC. Turn right (turning IAS MAX 250 kt) to intercept and follow magnetic track 076° SGO direct to cross NDB SGO at 4500 ft or above. Proceed on magnetic track 045° SGO direct to SOPET.
5.5% minimum climb gradient to 2000 ft.

3.2 Enroute Chart

- **Function** - the Enroute Chart shall provide the crew with information to facilitate navigation along ATS routes in compliance with ATS procedures.

- **Availability** - the Enroute Chart shall be available wherever a Flight Information Region (FIR) has been established.

- **Coverage and scale** - aware of the varying degree of congestion of information over certain areas, ICAO does not specify a uniform scale for Enroute Charts. However, it recommends to avoid large variations of scale between adjacent charts, and ensure an adequate overlap to facilitate the continuity of navigation. A linear scale based on the mean scale of the chart may be shown.

- **Projection** - conformal projections on which straight lines approximate great circle (orthodromic) trajectories should be used. Parallels and meridians shall be shown at suitable intervals.

- **Identification** - each sheet shall be identified by chart series and number (see Figure [12.4]).

- **Culture and topography** - generalized shore lines of all open water areas, lakes and rivers shall be shown unless they conflict with data more applicable to the function of the chart. Within each quadrilateral formed by parallels and meridians, the area of minimum altitude shall be shown. When charts are not orientated towards the True North, it must be clearly indicated, as well as the selected criteria.

- **Magnetic variation** - ICAO recommends to show isogonal lines[4] and the date of the given isogonic information.

- **Bearings, tracks and radials** shall be magnetic, except for areas of very high latitude, where it is determined by the appropiate authority that reference to Magnetic North is impractical. If any bearing, track or radial is given with respect to the True North, it must be clearly specified.

- **Aeronautical data**

  - All aerodromes used by international civil aviation to which an instrument approach can be made shall be shown.

  - Prohibited, restricted and danger areas, relevant to the layer of airspace shown on the chart will be depicted, with their identification and vertical limits.

---

[4]Isogonal lines join points with the same magnetic variation

- Indications of all designated airspace, including its class, lateral and vertical limits.

- Radio navigation aids associated with the ATS system, including their names, identifications, frequencies and geographical coordinates.

- ATS routes for enroute flight including route designators, required navigation performance (RNP) if applies, track and segment distances, as well as the direction of traffic flow.

- Minimum flight altitudes on ATS routes

- Radio communication facilities listed with their frequencies[5]

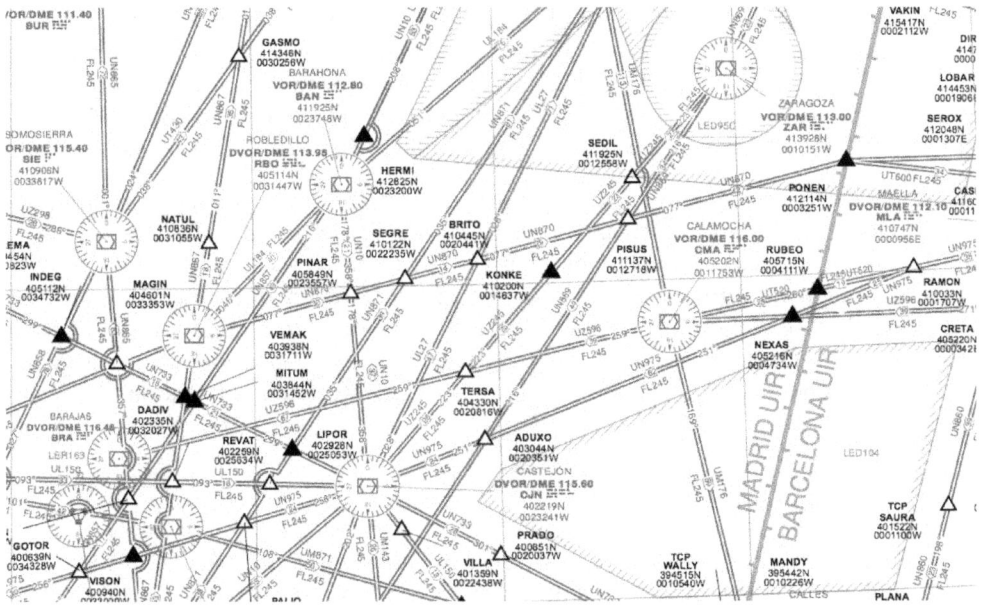

Figure 12.5: Upper airspace Enroute Chart for part of the spanish territory as published in the Spanish AIP © Enaire

3.3 Instrument Approach Chart (IAC)

- **Function** - the Instrument Approach Chart shall provide the crew with information to perform an approved instrument procedure to the runway of intended landing including the missed approach procedure and where applicable, the associated holding patterns.

---

[5]That is, frequencies associated to the sectors that provide the different ATS services within the depicted area

- **Availability** - the Instrument Approach Chart shall be available for all aerodromes used by international civil aviation where instrument approach procedures have been established. A separate IAC shall normally be provided for each precision and non-precision approach procedure.

- **Coverage and scale** - the coverage shall be sufficient to include all segments of the instrument approach procedure, with a scale that ensures optimum legibility. A circle with a radius of 10nm centered typically on the ARP[6] shall be shown.

- **Projection** - conformal projections on which straight lines approximate great circle (orthodromic) trajectories shall be used.

- **Identification** - the IAC shall be identified by the name of the city to which the aerodrome serves, the name of the aerodrome, the type of navaid on which the approach procedure relies and the runway designator. If operational restrictions such as approaches designed for certain types of aircraft, RNP or RNAV limitations apply, it must be established at the identification header.

- **Culture and topography** - the minimum shall be a delineation of land masses and significant lakes and rivers pertinent to the safe execution of the approach procedure, including the missed approach and any associated holding procedure or visual maneuvering (circling). Appropiate contour lines and important spot elevations shall be presented. Topography in the vicinity of the approach is of great importance to avoid possible CFIT, a good description of the environment and obstacles is thus of extreme importance to increase situational awareness.

- **Magnetic variation** should be shown.

- **Bearings, tracks and radials** shall be magnetic, except for areas of very high latitude, where it is determined by the appropiate authority that reference to Magnetic North is impractical. If any bearing, track or radial is given with respect to the True North, it must be clearly specified.

- **Aeronautical data**

  - Aerodromes in the vicinity (whose traffic patterns may affect the approach) and the aerodrome on which the procedure is based shall be clearly shown, including their runways properly orientated and their elevation or that of their runway thresholds.

  - The elevation of any obstacle affecting the maneuver above mean sea level. If heights above a datum different from mean sea level should be given, this should be in parentheses. The datum shall typically be the

---

[6]Airport Reference Point

threshold elevation of the runway to which the approach procedure is related.

— Prohibited, restricted or danger areas which may affect the execution of the procedures shall be shown with their identification and vertical limits.

— The FAF (Final Approach Fix) and other essential fixes or points comprising the procedure shall be shown and identified.

— Priority shall be given when plotting the procedure to those navaids that are necessary for its performance, including the names and frequencies of their ground stations.

— The Minimum Sector Altitude (MSA) established by the competent authority shall be shown, with a clear indication of the sector to which it applies.

— The approach procedure track shall be depicted by an arrowed continuous line indicating the direction of the flight. The missed approach procedure track by an arrowed broken line.

— Where necessary, the holding pattern and minimum holding altitude associated with the approach and missed approach.

— Aerodrome operating minima, together with obstacle clearance altitude / heights shall be shown.

In the following two pages, both a precision (ILS) and a non-precision (VOR) approach procedures are depicted for the same runway 30 of GCXO. Please, notice the difference between the definition of the vertical profiles.

CARTA DE APROXIMACIÓN
POR INSTRUMENTOS-OACI

© Enaire
NOT FOR OPERATIONAL USE

ELEV AD
2077

APP 124.800	TENERIFE NORTE
TWR 118.700	ILS
GMC 121.700	RWY 30
ATIS 118.575	

ALTITUDES, ELEVACIONES Y ALTURAS EN PIES.
DISTANCIAS EN MILLAS NÁUTICAS.
LAS MARCACIONES SON MAGNÉTICAS.

NOTA: SUJETA A LA ACTIVIDAD DE LA GCD29

DME REQUERIDO
DVOR REQUERIDO
ADF REQUERIDO

MSA
DVOR/DME TFN
6500
14 500
25 NM

NO EN CIRCUITO

(IAF)
TENERIFE NORTE
DVOR/DME 117.70
TFN
28°32'13"N
016°16'07"W

A DVOR/DME TFN

LOC 110.30
ITF H
2553 (778)

TENERIFE NORTE
L 420
FP
28°29'31"N
016°22'10"W

TENERIFE NORTE
DVOR/DME 116.20
LRO
28°29'05"N
016°21'06"W

3395 (1035)   *2986

GCD29 6000 ft ALT SEA

IAS MAX VIRAJE

ALABEO	15°	20°	25°
IAS MAX (kt)	185	210	230

R-117 TFN

2034 (361)

(FAP)
6.9 DME ILS
8.3 DME LRO
28°26'14"N
016°12'16"W

(IF)
12.8 DME ILS
14.2 DME LRO
12.0 DME TFN
28°24'08"N
016°06'00"W

IAS MAX 185 kt

R-128 TFN

GCD25 6000 ft ALT SEA

GCD74 FL330 SEA/GND

(IAF)
CANDE
24.0 DME TFN
28°19'25"N
015°53'05"W

MNM ALT FL70

29.0 DME TFN

ESCALA 1:400 000

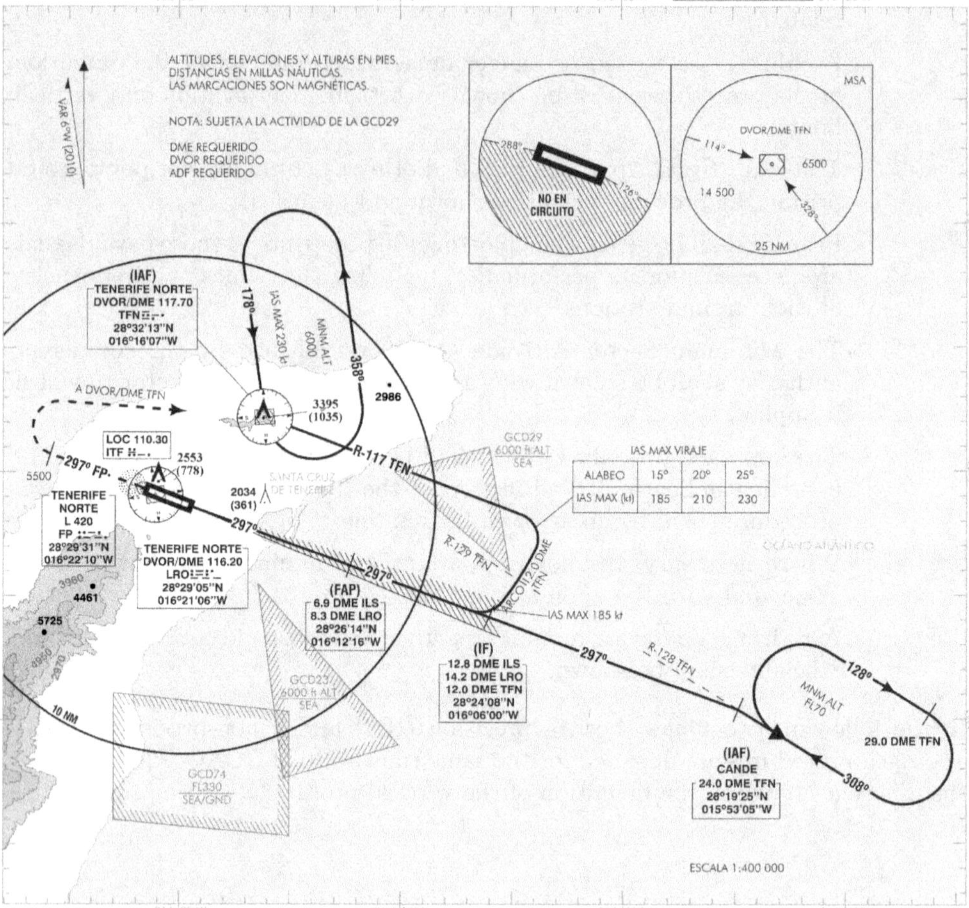

FRUSTRADA: SUBIR EN RUMBO DE PISTA DIRECTO A L FP, PROCEDER POR RUTA MAGNÉTICA 297° FP HASTA ALCANZAR 5500 ft. VIRAR A LA DERECHA DIRECTO AL DVOR/DME TFN.
INCORPORARSE A LA ESPERA A 6000 ft.

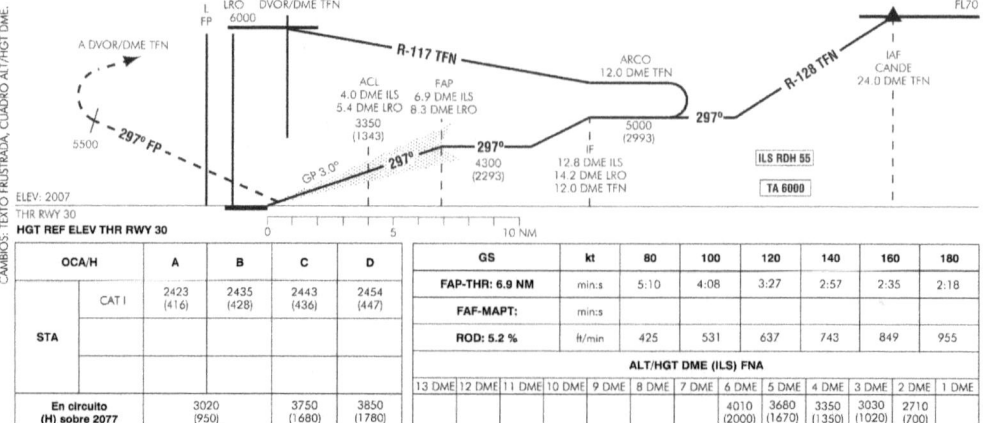

ELEV: 2007
THR RWY 30
HGT REF ELEV THR RWY 30

OCA/H	A	B	C	D
CAT I	2423 (416)	2435 (428)	2443 (436)	2454 (447)
STA				
En circuito (H) sobre 2077	3020 (950)		3750 (1680)	3850 (1780)

GS	kt	80	100	120	140	160	180
FAP-THR: 6.9 NM	min:s	5:10	4:08	3:27	2:57	2:35	2:18
FAF-MAPT:	min:s						
ROD: 5.2 %	ft/min	425	531	637	743	849	955

ALT/HGT DME (ILS) FNA

13 DME	12 DME	11 DME	10 DME	9 DME	8 DME	7 DME	6 DME	5 DME	4 DME	3 DME	2 DME	1 DME
							4010 (2000)	3680 (1670)	3350 (1350)	3030 (1020)	2710 (700)	

CAMBIOS: TEXTO FRUSTRADA, CUADRO ALT/HGT DME.

WEF 08-JAN-15 (AIRAC AMDT 15/14)   AIP-ESPAÑA   AD 2-GCXO IAC/3

CARTA DE APROXIMACIÓN POR INSTRUMENTOS-OACI

© Enaire

NOT FOR OPERATIONAL USE

ELEV AD 2077

APP	124.800
TWR	118.700
GMC	121.700
ATIS	118.575

TENERIFE NORTE
VOR
RWY 30

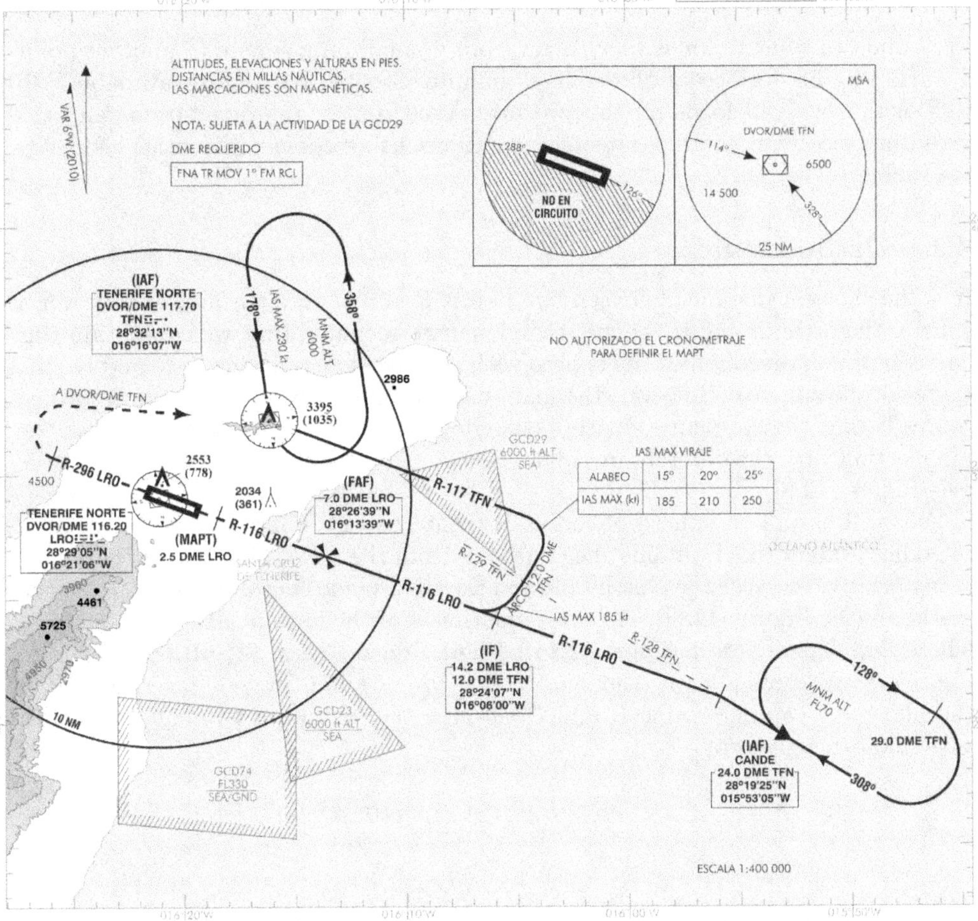

VAR 6° W (2010)

ALTITUDES, ELEVACIONES Y ALTURAS EN PIES.
DISTANCIAS EN MILLAS NÁUTICAS.
LAS MARCACIONES SON MAGNÉTICAS.

NOTA: SUJETA A LA ACTIVIDAD DE LA GCD29

DME REQUERIDO

FNA TR MOV 1° FM RCL

MSA

288°
NO EN CIRCUITO
126°
14 500

DVOR/DME TFN
114°
6500
338°
25 NM

(IAF)
TENERIFE NORTE
DVOR/DME 117.70
TFN
28°32'13"N
016°16'07"W

IAS MAX 230 kt

178°   358°

MNM ALT 6000

2986

NO AUTORIZADO EL CRONOMETRAJE
PARA DEFINIR EL MAPT

A DVOR/DME TFN

3395 (1035)

R-296 LRO   4500

2553 (778)

2034 (361)

TENERIFE NORTE
DVOR/DME 116.20
LRO
28°29'05"N
016°21'06"W

(MAPT)
2.5 DME LRO

R-116 LRO

(FAF)
7.0 DME LRO
28°26'39"N
016°13'39"W

SANTA CRUZ DE TENERIFE

R-117 TFN

GCD29
6000 ft ALT
SEA

IAS MAX VIRAJE			
ALABEO	15°	20°	25°
IAS MAX (kt)	185	210	250

R-129 TFN

ARCO 12.0 DME TFN

OCÉANO ATLÁNTICO

IAS MAX 185 kt

R-116 LRO

(IF)
14.2 DME LRO
12.0 DME TFN
28°24'07"N
016°06'00"W

R-116 LRO

R-128 TFN

128°

MNM ALT FL70

29.0 DME TFN

(IAF)
CANDE
24.0 DME TFN
28°19'25"N
015°53'05"W

308°

4461
5725

10 NM

GCD23
6000 ft ALT
SEA

GCD74
FL330
SEA/GND

ESCALA 1:400 000

FRUSTRADA: SUBIR EN R-116 LRO DIRECTO AL DVOR/DME LRO. PROCEDER POR R-296 LRO HASTA ALCANZAR 4500 ft. VIRAR A LA DERECHA DIRECTO AL DVOR/DME TFN. INCORPORARSE A LA ESPERA A 6000 ft.

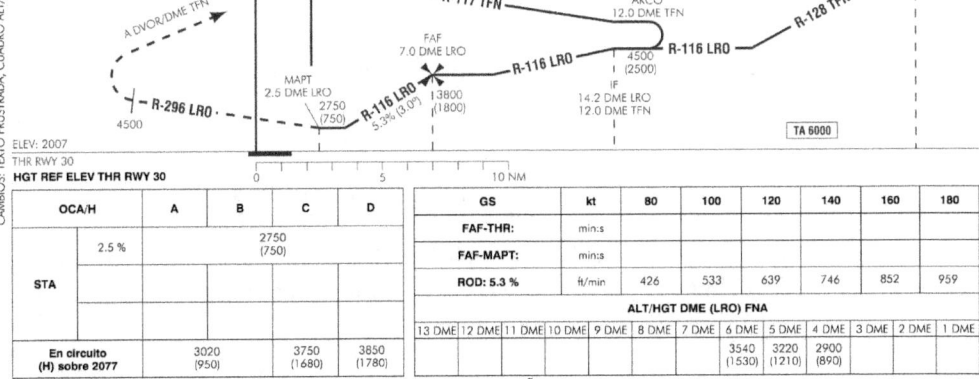

OCA/H	A	B	C	D
	2.5 %	2750 (750)		
STA				
En circuito (H) sobre 2077	3020 (950)		3750 (1680)	3850 (1780)

GS	kt	80	100	120	140	160	180
FAF-THR:	min:s						
FAF-MAPT:	min:s						
ROD: 5.3 %	ft/min	426	533	639	746	852	959

ALT/HGT DME (LRO) FNA													
13 DME	12 DME	11 DME	10 DME	9 DME	8 DME	7 DME	6 DME	5 DME	4 DME	3 DME	2 DME	1 DME	
							3540 (1530)	3220 (1210)	2900 (890)				

CAMBIOS: TEXTO FRUSTRADA, CUADRO ALT/HGT DME.

ELEV: 2007
THR RWY 30
HGT REF ELEV THR RWY 30

WEF 08-JAN-15 (AIRAC AMDT 15/14)   AIP-ESPAÑA   AD 2-GCXO IAC/5

4 Orthodrome and Loxodrome

Providing we want to know how to get from point P on a sphere to another point Q, also on the surface of the sphere, infinite possible curves are available. In this section we will focus on the two most used routes, the orthodrome and the loxodrome, solving their parametric equations using some differential geometry reasoning.

4.1 Orthodrome

It is the shortest distance between two points P and Q on a surface. When over a sphere, the orthodrome is the great circle curve, meaning that we can obtain this curve as the intersection of the sphere with a plane that contains P, Q and O (the center of the sphere). However, the main navigation inconvenience for orthodrome routes is that they are curves with a changing geographic track, meaning that the intersection with the different meridians is made at different angles along the curve (see Figure [12.8]).

We will assume a sphere of radius R, centered on O and with an associated reference system that contains the equator within the OXY plane, with the Z-axis being positive towards the North Pole and the OXZ plane containing the Greenwich meridian (see Figure [12.6]). A plane equation is of the form $x + By + Cz + D = 0$ where B, C and D are parameters to be determined. Since $O\,(0,0,0)$ belongs to the plane, we know $D = 0$. And since $P\,(x_P, y_P, z_P)$ and $Q\,(x_Q, y_Q, z_Q)$ belong to the plane, we can determine B and C through the system of two equations with two unknowns given by

$$x_P + B\,y_P + C\,z_P = 0$$

$$x_Q + B\,y_Q + C\,z_Q = 0$$

We know that the orthodrome will be on $x + By + Cz = 0$. Since the curve will also belong to the sphere, we know that $x = R\cos(\varphi)\cos(\lambda)$, $y = R\cos(\varphi)\sin(\lambda)$ and $z = R\sin(\varphi)$, meaning then that the following equation establishes a relationship between φ and λ.

$$cos(\varphi)\cos(\lambda) + B\,cos(\varphi)\,sin(\lambda) + C\,\,sin(\varphi) = 0$$

so

$$\varphi = \varphi(\lambda) = arctan\left\{\frac{-1}{C}\,[cos(\lambda) + B\,sin(\lambda)]\right\}$$

The shortest distance between P and Q can be determined using the dot product of $\vec{x_P}$ and $\vec{x_Q}$, being $\vec{x_P} \cdot \vec{x_Q} = R^2 \cdot cos(\theta)$, being θ the angle of the arc the

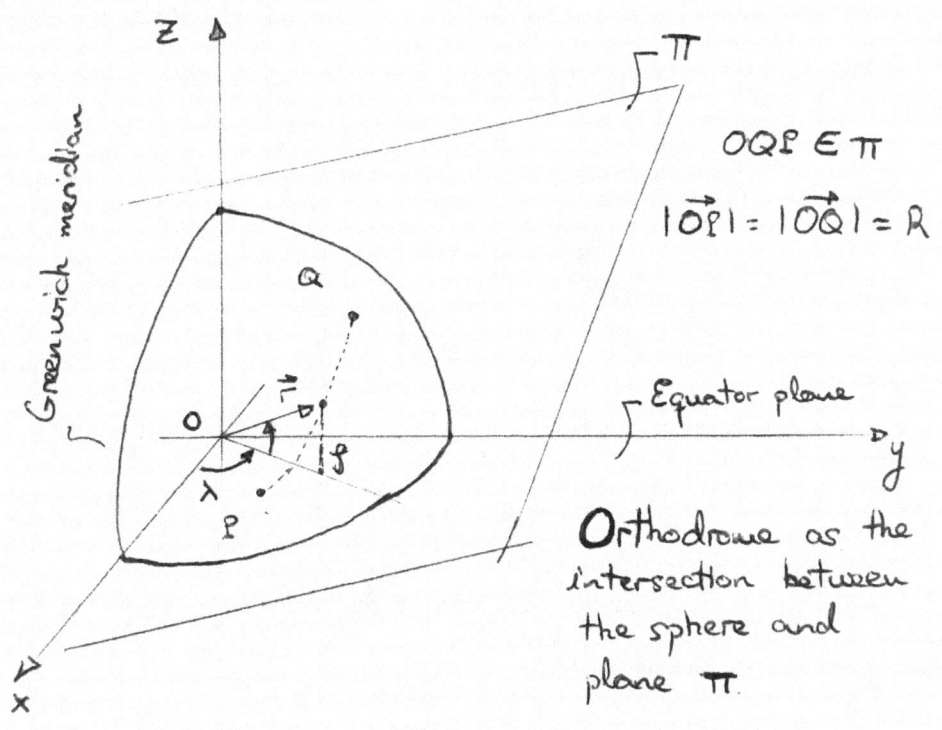

Figure 12.6: Detail of orthodrome geometry

orthodrome trajectory makes between P and Q. Solving for θ and later applying that the orthodrome distance $d = R \cdot \theta$, where R is the radius of the Earth.

The geographic track (γ) is mathematically defined by the dot product $\vec{t} \cdot \vec{u_\varphi} = cos(\gamma)$ where $\vec{t} = \frac{\vec{x}'}{\|\vec{x}'\|}$ is a unitary tangent vector in the direction of the curve, that can be obtained deriving and normalizing the parametric equations of the curve $\vec{x} = [x(\lambda), y(\lambda), z(\lambda)]$ with respect to its parameter λ, being $\vec{u_\varphi}$ a unitary vector on the surface of the Earth, in the direction of the meridian (positive northbound).

4.2 Loxodrome

Navigation is a lot simpler when track angle is constant. If we did so, all routes would end up being spirals converging in the North or South poles (unless we flew exact east or west routes). Loxodromes are curves that intersect meridians with a constant angle. The problem we want to solve here is, given two points, starting point $P(\varphi_P, \lambda_P)$ and ending point $Q(\varphi_Q, \lambda_Q)$, derive the parametric equation of the curve that intersects all meridians with a constant (and initially unknown) angle γ.

From Figure [12.7], through basic trigonometry for an infinitesimal triangle we know that

$$ds \cdot sin(\gamma) = r \cdot d\lambda = R \cdot cos(\varphi) \cdot d\lambda \tag{8}$$

$$ds \cdot cos(\gamma) = R \cdot d\varphi \tag{9}$$

dividing (8) by (9), we get

$$tan(\gamma) = \frac{cos(\varphi) \cdot d\lambda}{d\varphi} \tag{10}$$

which is a differential equation in which γ is constant and that can therefore be integrated:

$$\int_{\varphi_P}^{\varphi} \frac{d\varphi}{cos(\varphi)} = \int_{\lambda_P}^{\lambda} \frac{d\lambda}{tan(\gamma)} \tag{11}$$

In order to integrate equation (11) we must first make a couple of changes of variables, saying first $cos(\varphi) = t$ and later $z = sin(\varphi) = \sqrt{1 - cos^2(\varphi)}$. After doing some algebra, we get:

$$argtanh[sin(\varphi)] - argtanh[sin(\varphi_P)] = \frac{\lambda - \lambda_P}{tan(\gamma)} \tag{12}$$

And since $Q(\varphi_Q, \lambda_Q)$ belongs to the curve, we can get

$$tan(\gamma) = \frac{\lambda_Q - \lambda_P}{argtanh[sin(\varphi_Q)] - argtanh[sin(\varphi_P)]} \tag{13}$$

and thus the track angle for the loxodromic. Once γ is retrieved, the relationship $\varphi = \varphi(\lambda)$ and thus the curve comes determined by the equation

$$\varphi = \varphi(\lambda) = arcsin \left\{ tanh \left[argtanh(sin\varphi_P) + \frac{\lambda - \lambda_P}{tan(\gamma)} \right] \right\} \tag{14}$$

The total length of the loxodrome curve can be computed using basic differential geometry equations, being

$$L = \int_{s_P}^{s_Q} ds = \int_{s_P}^{s_Q} ||\overrightarrow{dr}|| = \int_{\lambda_P}^{\lambda_Q} ||\overrightarrow{x}'(\lambda)|| \cdot d\lambda \tag{15}$$

An example of a route between Madrid and Tokyo has been performed (see Figure [12.8]), showing clearly that a more simple navigation (loxodrome, constant geographic track) is longer in terms of distance relative to the orthodrome[7].

---

[7]6770 nautical miles for the loxodrome route versus 5804 nautical miles for the orthodrome

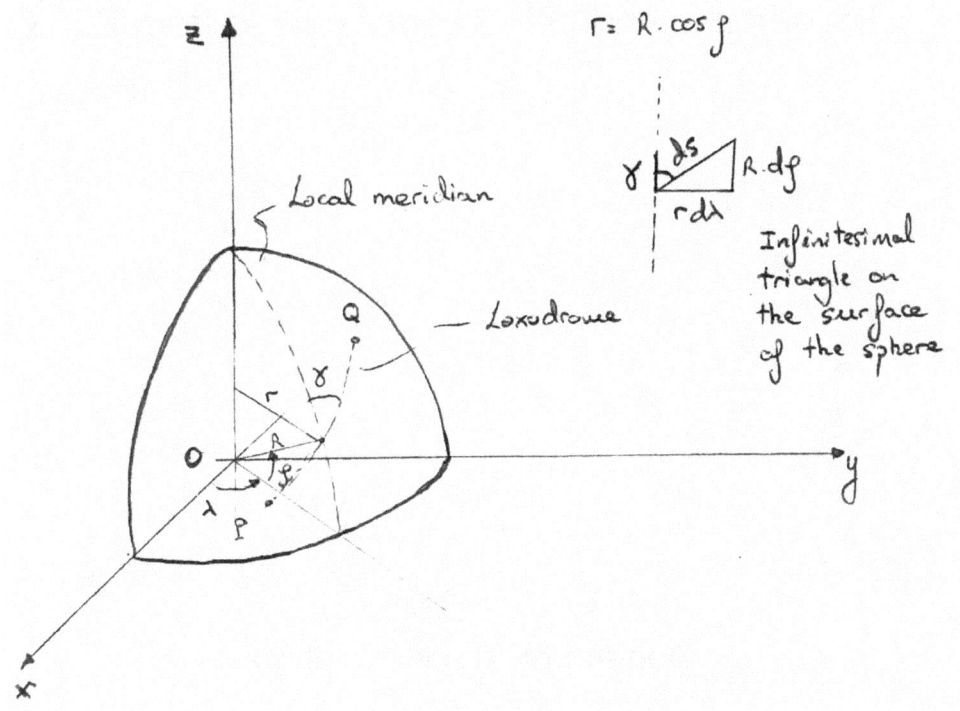

Figure 12.7: Detail of loxodrome geometry

However, the representation on a plane plot, shows the orthodrome latitude versus longitude evolution curved and might make the reader falsely believe that the route is longer.

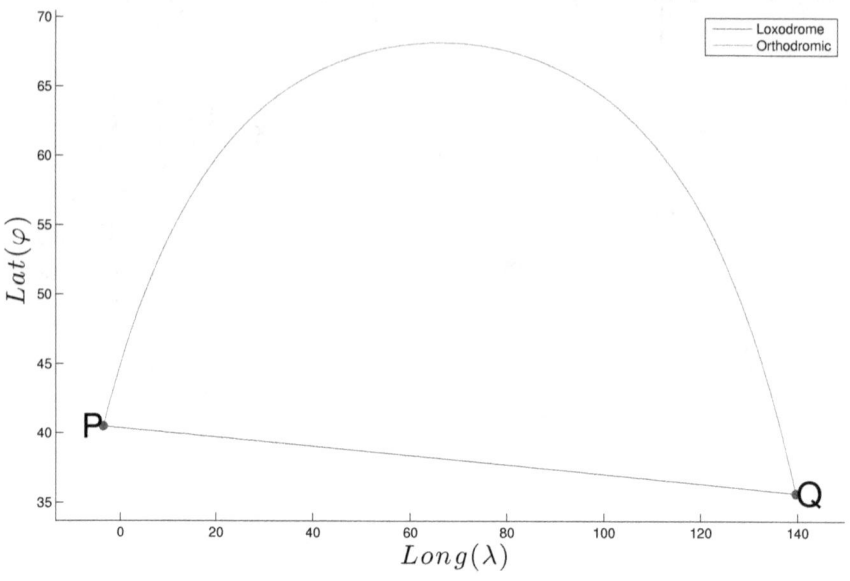

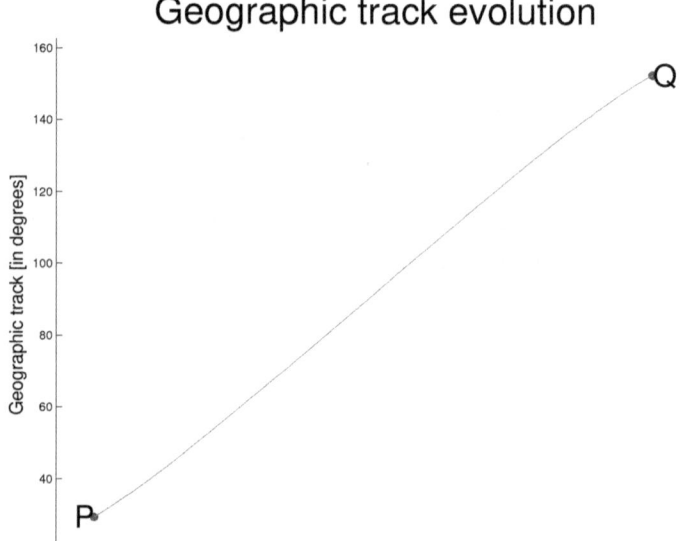

Figure 12.8: $\varphi = \varphi(\lambda)$ evolution for both routes and Geographic Track evolution for the orthodrome curve (constant $\gamma = 92.44^o$ for the Loxodrome) [Madrid (P) to Tokyo (Q)]

Chapter 13. Area Navigation

1 From Conventional to Area Navigation

Traditionally, ground navaids such as VORs, NDBs and DMEs have driven the positioning and navigation of aircraft. Pilots would, through their instruments (a Radio Magnetic Indicator, RMI, or a Course Deviation Indicator, CDI), know their position relative to these ground navaids, and could easily fly *to* or *from* them, but hardly towards any other geographic point. Navigation was thus limited by the range of ground navaids and the ability of the onboard equipment to monitor their signals, which typically ranged not further than 150 nm. This imposed limitations to navigation and aircraft were forced to follow ATS routes that had to be defined between the different installed ground navaids. The network of airways was thus defined depending on the availability of ground navaids and aircraft were forced to fly zig zag routes until reaching their destination, with the associated impact on efficiency. Figure [13.1] shows how Conventional Navigation is performed. These networks of routes are still in use today, and are typically defined in Lower Airspaces (below FL245), as can be seen in Figure [13.2], where all ATS routes are defined based on the availability of certain navaids.

Navigating with respect to traditional ground navaids using traditional equipment allows pilots not only to fly straight magnetic routes with respect to the monitored navaid, but also arcs (defined at given DME distances from the navaid), which are commonly used for the definition of conventional standard instrument departures and arrivals (SIDs & STARs). See Figure [13.3].

The inability to fly direct routes impacts flight times and wasted fuel. With conventional equipment onboard, pilots can hardly fly direct routes between random waypoints, even if defined with respect to the same navaid. Trigonometry computations would be required to define any random new track, adding up the latter difficulty to monitor trajectory adherence, not to mention the practical inability to constantly monitor the influence of the wind in the trajectory. Point to point navigation using conventional instruments requires an important amount of manual computations plus the use of dead reckoning techniques, being unable to

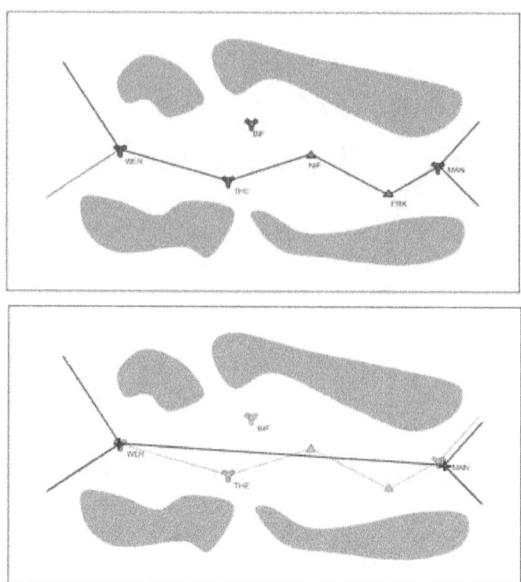

Figure 13.1: Conventional vs Area Navigation © ICAO Doc 9613. *Reproduced with the permission of ICAO*

constantly correct for crosswind components. This would add a lot of workload to the pilots and could lead to unacceptable levels of safety. It was thus not until computers were brought onboard, facilitating these computations, that Conventional Navigation gave way to a more efficient Area Navigation (RNAV).

Area Navigation (RNAV) is a method of navigation that enables the lateral navigation of an aircraft in any desired path. In fact, ICAO defines RNAV as *a method of navigation which permits aircraft operation on any desired flight path within the coverage of ground-based or space-based navigation aids or within the limits of self-contained aids, or a combination of these.* Unlike in Conventional Navigation, in RNAV an aircraft can determine its position and navigate independently on whether it is or it is not on any radial or track with respect to any ground navaid. In the early 1970s first digital avionics appeared and were installed on commercial aircraft. One of the pioneers was the Lockheed L1011 Tristar, where pilots were able to introduce up to four different waypoints defined by coordinates to support a CDI and navigate the route. From then on, computers started being installed onboard to support navigation. The storage capacity surged, increasing the ability to have a complete database of waypoints. New avionics enabled the appearance of the first Navigation Displays, which showed a clearer picture of the aircraft´s position and its navigation intentions. By the 1990s, increasing levels of traffic made some ground navaids experience saturation (due to an excess in the

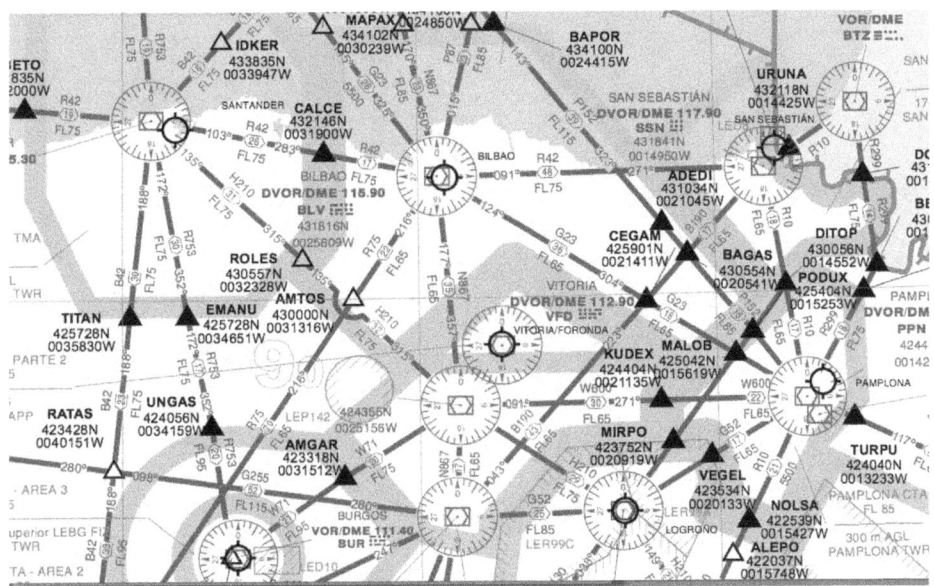

Figure 13.2: Low Airspace definition enabling Conventional Navigation © Enaire

number of interrogations from aircraft). Because of these reasons, European authorities made a move in April 1998 declaring all Upper European Airspace (above FL245) should have an RNAV Enroute Structure. All upper ATS routes from then on are not defined based on the existence and availability of ground navaids, but on other facts such as track efficiency, the existence of restricted areas, demand requirements or sectorization and ATM reasons.

Although it may sound obvious, the implementation of an RNAV ATS route structure required that these ATS routes could only be flown by aircraft capable of employing Area Navigation, that is, only aircraft equipped with *RNAV systems* can effectively fly these routes.

2 RNAV Systems

The main goal of an RNAV System (installed onboard), is to generate steering commands that make an aircraft navigate by comparing its estimated position with the defined lateral route to be followed.

2.1 Determination of position

The determination of the position is done through the use of different aircraft systems, including as many ground- and space-based navaids as possible.

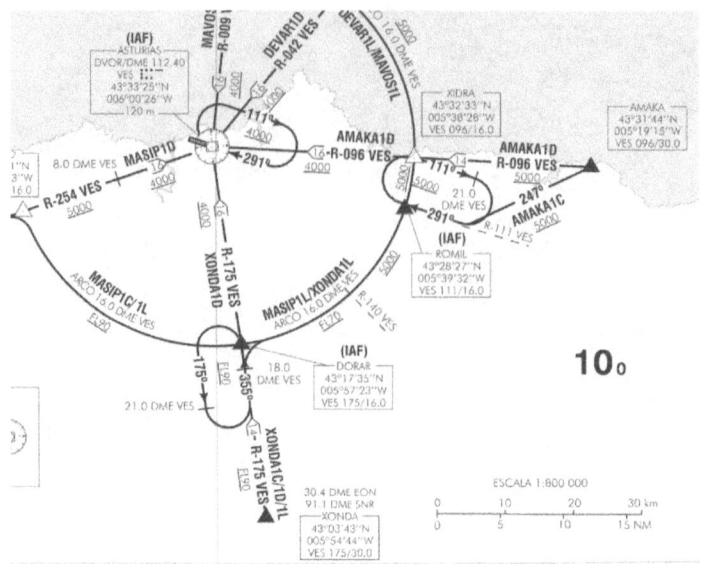

Figure 13.3: Conventional STAR Procedure including a DME arc © Enaire

- **VOR/DME** or **DME/DME** equipment can be used to determine the relative position of the aircraft with respect to a set of ground navaids (e.g. radial and distance to a given VOR, or intersection of DME arcs to a couple of stations) whose global positions in turn are stored in the onboard computers, that can thus determine the global position of the aircraft (adding up positioning vectors, using geographic coordinates).

- **GNSS** (Global Navigation Satellite System) is the common facilitator under which all current and future navigation developments are expected to happen. Different satellite constelations are included under the GNSS acronym, being the most important the american GPS, the russian GLONASS, the chinese COMPASS and the european GALILEO. Availability and quality of the signals received from each one of these global constelations drive the accuracy in the determination of the receiver´s position. There are different methods to enhance their performance, as we will see in section 4.

- **IRS** (Inertial Reference System) is an onboard totally autonomous system that outputs (among others) the latitude and longitude of the aircraft, based on the accelerations suffered by the aircraft in time. The accuracy of IRS outputs deteriorates with time, requiring periodic input from external sources to update for errors in the determination of position. Honeywell is a well-known manufacturer of IRS equipments.

2.2   Database

Onboard computers need a database which includes a set of *waypoints*. These are referred to a global reference system. This is necessary in order to define the intended path and compare the retrieved position of the aircraft with the desired trajectory (being thus able to output the required steering commands). Onboard navigation systems use an ECEF (Earth Centered Earth Fixed) reference system, longitude (λ) and latitude (φ) as described in the chapter devoted to cartography.

ICAO defines a **waypoint** as *a specified geographical location used to define an area navigation route of the flight path of an aircraft employing area navigation.* They are identified with a unique name code of up to 5 letters (including vowels, making it easy to pronounce). Only if they are defined in a Terminal Area (for the use of a specific departure or arrival procedure), they are named with a couple of letters refering to the airport to which they are serving and an identification number (e.g. MD002). If a waypoint is defined over a navaid, it may have the name of the navaid (3 letters typically for a VOR, and 2 or 3 letters for a NDB).

Waypoints are not only defined by their geographic coordinates, they are also characterized by the ATS route to which they belong and by the manner in which they have to be flown:

- **Fly-by** waypoints are those that should not be overflown. Instead, the aircraft must initiate the turn before reaching the waypoint, in order to intercept the outbound flight leg in a tangent trajectory. They are normally defined within Enroute ATS routes.

- **Fly-over** waypoints are those that have to be completely overflown (no anticipation in the turn is allowed). They are typically defined in Terminal Area procedures.

Whether a given waypoint is defined as Fly-by or Fly-over has to appear clearly on the navigation charts. Figure [13.4] illustrates the different cartography coding for each waypoint type.

Onboard systems are normally equipped with a database that includes all worldwide waypoints, ATS routes, instrument procedures and associated restrictions (similarly to what can be observed in Figure [13.5]). It is of extreme importance that both, aircraft operators and air navigation support facilities (in the tactical phase being the actors namely pilots and air traffic controllers), have the same database. Because of this, the AIRAC Cycle is defined, which is a 28-days-valid database. Any change in the definition of waypoints, procedures, ATS routes, etc, has to be implemented with a new AIRAC Cycle.

2.3   Errors

Due to different factors, the path defined by the RNAV System may not match exactly the desired path, and the estimated position of the aircraft by the different systems (see section 2.1) may not match the actual position.

Navegación de área / Área navigation				
Punto / Point	De paso / Fly-by		De sobrevuelo / Fly-over	
Notificación Reporting	No obligatoria On request	Obligatoria Compulsory	No obligatoria On request	Obligatoria Compulsory
VORTAC				
TACAN				
VOR				
VOR/DME				
NDB				
Punto de recorrido / Way- point				

Figure 13.4: Different waypoint representation on the Spanish AIP © Enaire

The following errors (see Figure [13.6]) are thus defined:

- Navigation System Error (NSE) - refers to the difference between the estimated and the actual position of the aircraft (a.k.a. Position Estimation Error)

- Flight Technical Error (FTE) - relates to the crew or autopilot ability to fly along the defined path (a.k.a. Path Steering Error)

- Path Definition Error (PDE) - reflects computational, display and navigation database errors in the RNAV System

The Total System Error (TSE) is used for PBN operations, as we will later see. It is defined $TSE = \sqrt{NSE^2 + FTE^2 + PDE^2}$.

3 Performance Based Navigation (PBN)

PBN is defined by ICAO in Doc 9613 as *Area Navigation based on performance requirements for aircraft operating along an ATS route, on an instrument approach procedure or in a designated airspace.* Being these requirements those that set clearly the minimum acceptable values of integrity, availability and accuracy for different PBN related magnitudes that affect the determination of position and the adherence of the navigation.

From the 1970s and until the first decade of the 21st century, the development of RNAV techniques was made in such a way that each region adopted different strategies to regulate and establish the minimum required standards for its implementation (ranging from avionics or crew procedures to airspace design). ICAO identified this lack of harmonization as a problem and launched the PBN concept in 2007, looking forward to meeting common needs and defining global standards.

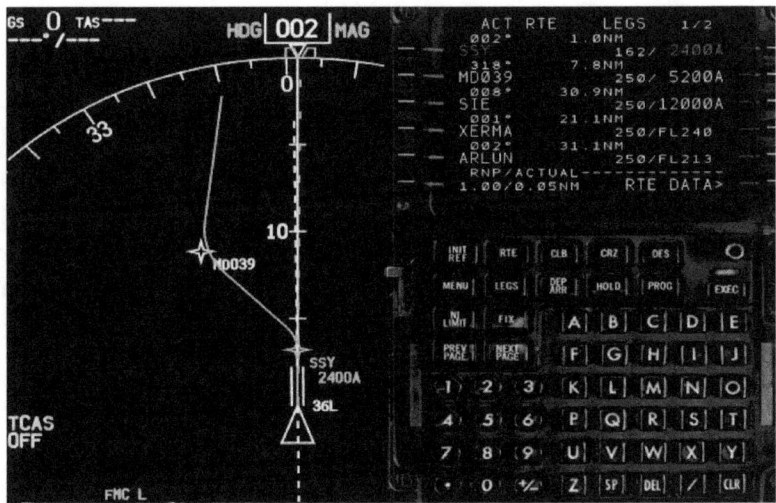

Figure 13.5: Representation of waypoints on a ND and associated trajectory definition as introduced in the Flight Management Computer

This new approach defines the different airspace and ATS route requirements (e.g. required TSE) as a function of the necessary aircraft performance, instead of worrying about the sources of the information or the employed methods. PBN cares more about the obtained result of a positioning or navigation process than about the means used to obtain these results.

Within this new context Conventional Navigation has no room at all. PBN is totally supported by Area Navigation. Applications that do not require Area Navigation are not part of the PBN concept.

If traditionally the Airspace Concept had been based on the application of each of the three pillars of CNS (Communication, Navigation & Surveillance), the irruption of RNAV makes us rethink about the Navigation pillar, splitting it into two new subpillars, namely the Performance Based Navigation (PBN), solving for each individual aircraft, and the Air Traffic Management (ATM), looking at the system as a complex network.

The implementation of the PBN concept to solve for a given airspace requires the definition of a navaid infrastructure, a navigation specification and application.

3.1 Navaid Infrastructure

Navaid Infrastructure refers to the ground- or space-based navigation aids. When defining the PBN requirements for a given airspace, each State will have to consider the navaid infrastructure to be deployed, together with the navaid policy. States should decide which, where and when it is worth it to deploy or remove certain

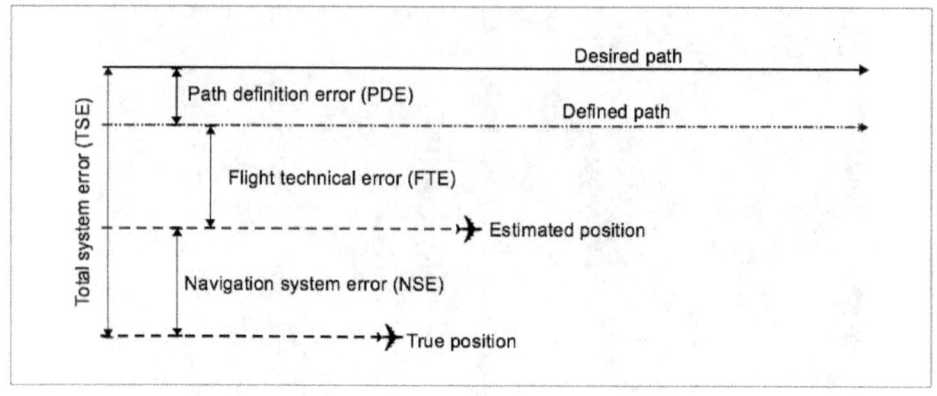

Figure 13.6: RNAV System errors © ICAO Doc 9613. *Reproduced with the permission of ICAO*

ground-navaids, depending on the expected trade-off between safety, efficiency and cost.

3.2 Navigation Specification

The Navigation Specification is defined by ICAO as *a set of aircraft and aircrew requirements needed to support PBN operations within a defined airspace*. It refers, among others, to the need to constantly monitor the accuracy and integrity in the estimation of navigation errors along the flight, and to the maximum required values for these errors. Depending on whether it is mandatory or not to monitor the Total System Error (TSE), we may define two kinds of Navigation Specification:

- RNAV Specification - does not include the requirement for performance monitoring and alerting. It is designated by the prefix RNAV, e.g. RNAV 5, RNAV 1. Not to be mistaken with Area Navigation (RNAV).

- RNP Specification - includes the requirement for performance monitoring and alerting, designated by the prefix RNP, e.g. RNP 4, RNP APCH, RNP (AR) APCH.

The number in the designator represents the minimum lateral navigation accuracy that has to be maintained at least for 95% of the time, limiting thus the required Total System Error (TSE). However, a specification is not only defined by the establishment of a TSE limit and the need (or not) for a monitoring and alerting system. Other requirements may also apply. For instance, RNAV 5 specification does not require a complete onboard database, while the more complex RNP (AR) APCH requires the ability to follow fixed radius turns. Also, complying with RNP specifications does not imply complying with RNAV specifications since there may be different equipment, training of contingency requirements. Each navigation specification has its own requirements, and no matter how tough these requirements seem to be, we should not give it for granted that another apparently "less restrictive" specification is included (e.g. complying with specification RNP 1 does not automatically mean we are complying with specification RNP 4).

During a flight, an aircraft will go through different airspaces, each of which requiring a different navigation specification. For this reason, and given the complexity of each individual navigation specification, ICAO has created the A-RNP specification (Advanced RNP), which gathers RNAV 5, RNAV 2, RNAV 1, RNP 2, RNP 1 and RNP APCH.

The available navaid infrastructure may place restrictions to the ability to obtain certain navigation specifications.

ICAO Doc 9613, Volume II, contains further detailed guidance on navigation specifications.

We will now make some short comments[1] on the most used navigation specifications:

- RNAV 10 - it is broadly used in oceanic and remote areas, not requiring any ground-based navaid infrastructure. It is being substituted slowly by RNP 4. The minimum route spacing where RNAV 10 is used is 50 nm.

- RNAV 5 (or BRNAV) - it is commonly used in enroute environments. Currently all european upper airspace is BRNAV. It might be substituted in the medium-term future by RNP 2.

- RNAV 1 & RNAV 2 - they both integrate the old PRNAV, commonly used for the definition of SID or STAR procedures.

Current european regulation proposals look for the application of RNP 1 in TMA environments and RNP 2 in Enroute environments, which would allow the definition of fixed-radius-turn ATS routes, enabling a narrowing in the distance between parallel routes plus the use of lateral offsets.

3.3 Navigation Application

A Navigation Application is *the application of a Navigation Specification in the context of a Navaid Infrastructure on a specific ATS Route, procedure or airspace volume.*

Typical PBN Navigation applications will range from PBN defined ATS routes, to SIDs, STARs or approach procedures.

All PBN approach procedures have a RNP specification (they are never allowed under RNAV specifications). Once again, it is necessary to stress the difference between RNAV specifications and Area Navigation (RNAV) as a method used for navigation. We will find Navigation Applications defined for the approach phase

---

[1]The author wants to thank Rafael Pecos for his help in this topic

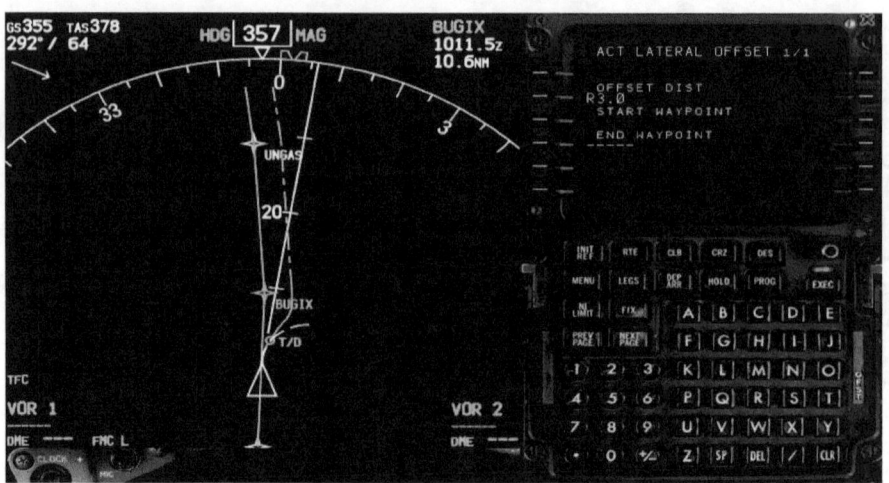

Figure 13.7: Lateral Offset of 3 nm depicted on a Navigation Display

that are RNAV based (Area Navigation is the used method), but never approach Navigation Applications that are RNAV specified.

The following chart shows an example of an RNAV precision approach procedure (where lateral and vertical guidance is obtained through GNSS augmented using SBAS) with a RNP APCH specification.

CARTA DE APROXIMACIÓN
POR INSTRUMENTOS-OACI

ELEV AD
16

APP 118.375
TWR 118.100
GMC 121.700

SANTANDER/Seve Ballesteros-Santander
RNAV (GNSS) Z
RWY 29

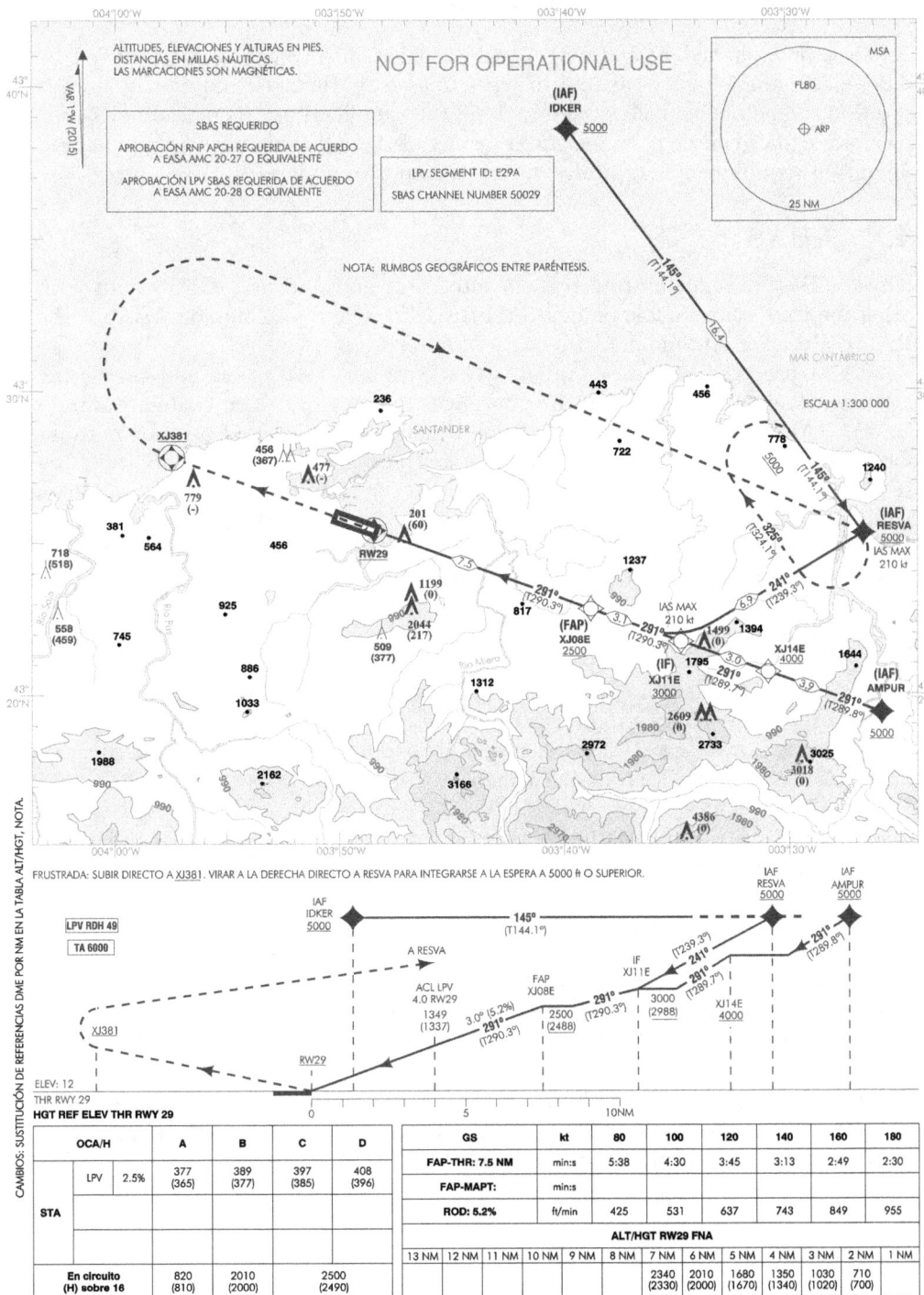

ALTITUDES, ELEVACIONES Y ALTURAS EN PIES.
DISTANCIAS EN MILLAS NÁUTICAS.
LAS MARCACIONES SON MAGNÉTICAS.

NOT FOR OPERATIONAL USE

SBAS REQUERIDO

APROBACIÓN RNP APCH REQUERIDA DE ACUERDO
A EASA AMC 20-27 O EQUIVALENTE

APROBACIÓN LPV SBAS REQUERIDA DE ACUERDO
A EASA AMC 20-28 O EQUIVALENTE

LPV SEGMENT ID: E29A

SBAS CHANNEL NUMBER 50029

NOTA: RUMBOS GEOGRÁFICOS ENTRE PARÉNTESIS.

MSA

FL80

⊕ ARP

25 NM

ESCALA 1:300 000

FRUSTRADA: SUBIR DIRECTO A XJ381. VIRAR A LA DERECHA DIRECTO A RESVA PARA INTEGRARSE A LA ESPERA A 5000 ft O SUPERIOR.

LPV RDH 49

TA 6000

ELEV: 12
THR RWY 29

HGT REF ELEV THR RWY 29

OCA/H		A	B	C	D
LPV	2.5%	377 (365)	389 (377)	397 (385)	408 (396)
STA					
En circuito (H) sobre 16		820 (810)	2010 (2000)	2500 (2490)	

GS		kt	80	100	120	140	160	180
FAP-THR: 7.5 NM		min:s	5:38	4:30	3:45	3:13	2:49	2:30
FAP-MAPT:		min:s						
ROD: 5.2%		ft/min	425	531	637	743	849	955

ALT/HGT RW29 FNA												
13 NM	12 NM	11 NM	10 NM	9 NM	8 NM	7 NM	6 NM	5 NM	4 NM	3 NM	2 NM	1 NM
						2340 (2330)	2010 (2000)	1680 (1670)	1350 (1340)	1030 (1020)	710 (700)	

4 Enhancing GNSS signals

As already explained in section 2.1, different satellite constelations powered by different economic regions are out in space to assist in the determination of aircraft position. Some of them share civil and military applications. The accuracy of the received signals can be complemented by any of the following means for enhanced operation requirements (typically during the approach phase).

4.1 ABAS

Aircraft Based Augmentation System integrates and improves GNSS data with other information available on board the aircraft. The most common ABAS is the RAIM (Receiver Autonomous Integrity Monitoring).

RAIM processes as many available GNSS signals as possible and checks the consistency among the different pseudoranges (when more than 4 satellites are in sight). RAIM complements the accuracy in the determination of position with information gathered from onboard barometers (baro-aided). The availability of the RAIM on each route or world area depends on the amount of operating satellites in sight. When there is an expected degraded level of RAIM, authorities must report through NOTAM publication. Inappropiate RAIM availability may lead to the inability to meet the required navigation performance.

4.2 SBAS

Besides the global satellite constellations that are used in the determination of position (GNSS primary satellites), other regional satellite constellations (SBAS GEO) are deployed (geostationary, providing service to the regions underneath). SBAS stands for Satellite Based Augmentation System.

A set of ground stations, whose real position is very well known, get signals from a GNSS primary constellation that let them, by comparison, estimate different GNSS primary errors. This information is treated in a grounded Processing Facility Center, computing for integrity, corrections and GEO ranging. The retrieved corrections are broadcasted in the shape of a message named Signal-In-Space (SIS) from the ground Processing Facility Center to all the SBAS GEO satellites, that relay these messages to the users. The SIS includes corrections for errors due to GNSS primary positioning, satellite clock/time deviations and ionospheric effects. EGNOS in Europe and WAAS in North American are the most extended SBAS GEO constellations with three SBAS GEO satellites each.

SBAS is currently used only during the approach phase (for maneuvers such as the one previously depicted for Santander Airport), although it is expected to be soon applied in enroute environments.

4.3 GBAS

Ground Based Augmentation Systems radiate, through VHF channels, differential corrections of GNSS primary constellations, letting the user have better integrity and accuracy of the signal. GBAS is typically used in the vicinity of airports, where ground antennas, whose position is very well known, are allocated. Range of the improved accuracy normally extends up to 30 nm from the airport, making it suitable only for the use of PBN approaches. Precision CAT I approaches can be constructed using this technology, being an ideal tool to complement or (even) substitute the traditional ILS. The following chart presents a GBAS based CAT I approach.

CARTA DE APROXIMACIÓN
POR INSTRUMENTOS-OACI

ELEV AD
52

MÁLAGA/Costa del Sol
GBAS Z
RWY 13

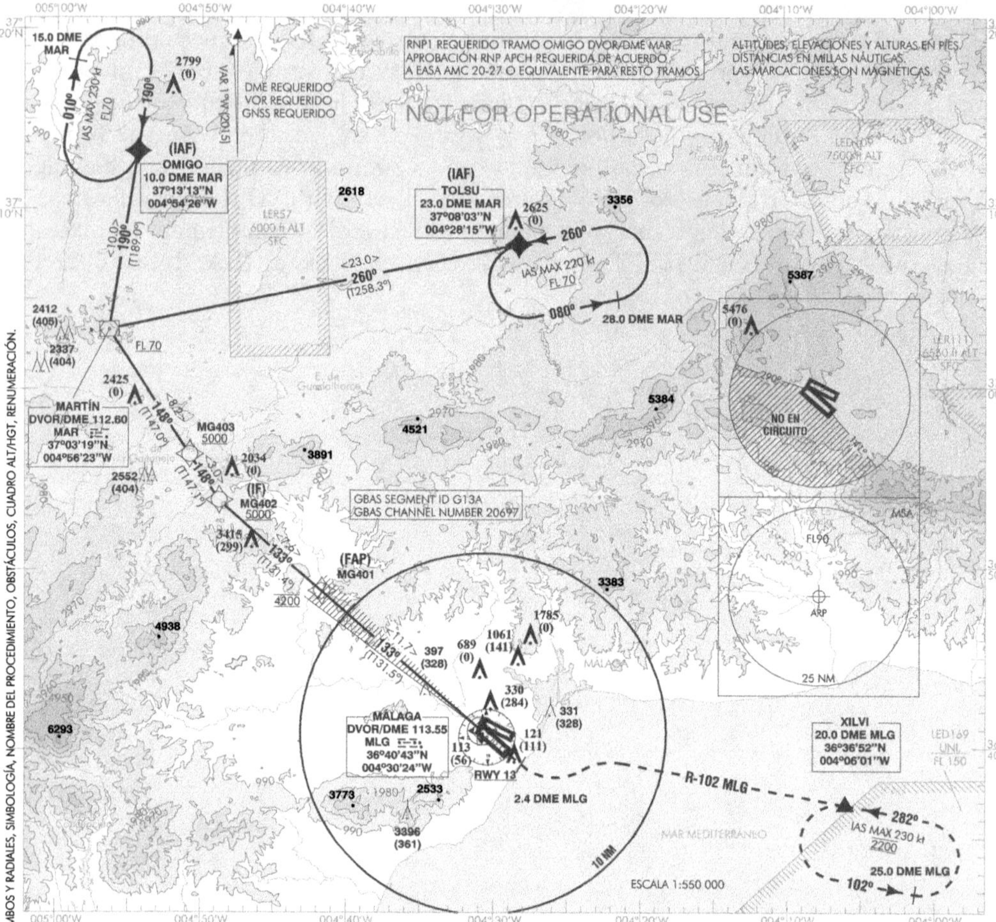

ARR	123.850
TWR W	118.150
GMC E	121.950
GMC W	121.700
ATIS ARR	120.375

NOT FOR OPERATIONAL USE

RNP1 REQUERIDO TRAMO OMIGO DVOR/DME MAR.
APROBACIÓN RNP APCH REQUERIDA DE ACUERDO
A EASA AMC 20-27 O EQUIVALENTE PARA RESTO TRAMOS

DME REQUERIDO
VOR REQUERIDO
GNSS REQUERIDO

ALTITUDES, ELEVACIONES Y ALTURAS EN PIES.
DISTANCIAS EN MILLAS NÁUTICAS.
LAS MARCACIONES SON MAGNÉTICAS.

15.0 DME MAR

010°
190°

2799 (0)

VAR 2°W (2015)

(IAF)
OMIGO
10.0 DME MAR
37°13'13"N
004°54'26"W

2618

(IAF)
TOLSU
23.0 DME MAR
37°08'03"N
004°28'15"W

2625 (0)

3358

260°

IAS MAX 220 kt
FL 70

080°
28.0 DME MAR

5387

5476 (0)

<0.0>
190°
(T189.0°)

<23.0>
260°
(T258.3°)

2412 (408)
2337 (404)

FL 70

5384

NO EN CIRCUITO

2425 (0)

MARTÍN
DVOR/DME 112.60
MAR
37°03'19"N
004°56'23"W

MG403 5000

148°

2034 (0)

3891

4521

GBAS SEGMENT ID G13A
GBAS CHANNEL NUMBER 20697

2552 (404)

(IF)
MG402 5000

148°

3415 (299)

133°

(FAP)
MG401

FL90

ARP

25 NM

4938

4200

133°

397 (131.5°)

689 (328)

1061 (141)

1785 (0)

330 (284)

331 (328)

6293

MÁLAGA
DVOR/DME 113.55
MLG
36°40'43"N
004°30'24"W

113 (56)

121 (111)

RWY 13

2.4 DME MLG

R-102 MLG

XILVI
20.0 DME MLG
36°36'52"N
004°06'01"W

LED169
UNL
FL 150

3773

2533

3396 (361)

MAR MEDITERRÁNEO

282°
IAS MAX 230 kt
2200

25.0 DME MLG

102°

ESCALA 1:550 000

FRUSTRADA: SUBIR EN RUMBO DE PISTA HASTA ALCANZAR 2.4 DME MLG. VIRAR A LA IZQUIERDA PARA INTERCEPTAR Y SEGUIR R-102 MLG HASTA XILVI E INTEGRARSE A LA ESPERA A 2200 ft.
ALTITUD MÁXIMA 2200 ft DURANTE LA MANIOBRA DE FRUSTRADA. ESPERAR INSTRUCCIONES ATC.
FRUSTRADA FALLO DE COMUNICACIONES: SUBIR EN RUMBO DE PISTA HASTA ALCANZAR 2.4 DME MLG. VIRAR A LA IZQUIERDA PARA INTERCEPTAR Y SEGUIR R-102 MLG PARA SOBREVOLAR
XILVI A 2200 ft. SEGUIR R-102 MLG HASTA 23.0 DME MLG. VIRAR A LA IZQUIERDA PARA SEGUIR ARCO 25.0 DME MLG HASTA R-104 MGA. VIRAR A LA IZQUIERDA PARA SEGUIR R-104 MGA
DIRECTO A DVOR/DME MGA FL70 PARA INCORPORARSE A LA LLEGADA PEKOP1Q.

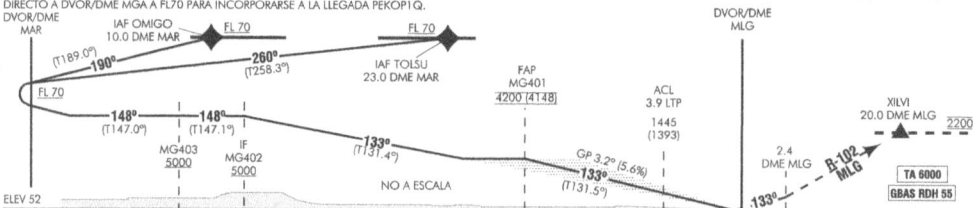

DVOR/DME
MAR

IAF OMIGO
10.0 DME MAR   FL 70

IAF TOLSU
23.0 DME MAR

FAP
MG401
4200 (4148)

ACL
3.9 LTP

DVOR/DME
MLG

XILVI
20.0 DME MLG   2200

190°
(T189.0°)

260°
(T258.3°)

FL 70

148°
(T147.0°)

148°
(T147.1°)

MG403 5000

IF
MG402 5000

133°
(T131.4°)

133°
(T131.5°)

1445 (1393)

GP 3.2° (5.6%)

2.4 DME MLG

R-102 MLG

TA 6000

GBAS RDH 55

ELEV 52

THR RWY 13

133°

GBAS RDH 55

NO A ESCALA

HGT REF ELEV THR RWY 13

OCA/H		A	B	C	D
	CAT I 2.5%	275 (223)	285 (233)	295 (243)	305 (253)
STA					
En circuito (H) sobre 52		650 (600)	1040 (990)	1470 (1420)	2200 (2150)

GS		kt	80	100	120	140	160	180
FAP-THR: 11.7 NM	min:s		8:46	7:01	5:51	5:01	4:23	3:54
FAF-MAPT:	min:s							
ROD: 5.6 %	ft/min		453	566	679	793	906	1019

	ALT/HGT DME THR13												
13 DME	12 DME	11 DME	10 DME	9 DME	8 DME	7 DME	6 DME	5 DME	4 DME	3 DME	2 DME	1 DME
		3960 (3900)	3600 (3550)	3240 (3190)	2890 (2830)	2530 (2480)	2180 (2130)	1830 (1780)	1480 (1430)	1140 (1090)	790 (740)	450 (400)

CAMBIOS: DECLINACIÓN MAGNÉTICA, ACTUALIZACIÓN DE RUMBOS Y RADIALES, SIMBOLOGÍA, NOMBRE DEL PROCEDIMIENTO, OBSTÁCULOS, CUADRO ALT/HGT, RENUMERACIÓN.

Chapter 14. IFR Procedures

In the Airspace Management chapter we already talked about the different IFR route structures. There we saw that for the enroute phase lower and upper IFR routes are defined. The ATS route network determines the route legs (defined between pairs of waypoints) that aircraft can choose when filing a flight plan. But, how do these network structures connect with the different airports? it is necessary to design procedures that connect enroute airways with airports, both for the departing and arriving flight phases. ICAO Doc 8168 (Volumes I and II) addresses these topics.

MOC (Minimum Obstacle Clearance) is defined by ICAO as the minimum altitude for a defined segment that provides the required obstacle clearance. Each leg of a published maneuver is designed taking into account the prescribed MOC values which are mandatory in compliance with ICAO Doc 8168. For *mountainous areas*, that is, areas of changing terrain profile where the changes of terrain elevation exceed 900 m (3000 ft) within a distance of less than 18.5 km (10.0 NM), the required MOC is typically doubled for safety purposes.

1 Standard Instrument Departures

A Standard Instrument Departure (SID) is defined by ICAO as *a designated IFR departure route linking the aerodrome or a specified runway of the aerodrome with a specified significant point, normally on a designated ATS route, at which the enroute phase of a flight commences.*

It is thus the purpose of a SID to connect an aerodrome with an ATS route through the definition of a procedure whose main concern will be the avoidance of terrain encounters (it must be designed in such a way that it provides obstacle clearance). Other secondary concerns such as the efficiency of the designed trajectory or noise abatement over certain populations may also be taken into account.

SIDs are named with the name of the last SID point (e.g. a waypoint or a navaid), a version number and a letter that differs depending on the runway of departure (e.g. CENTA7A departure from LEVC runway 30, indicating CENTA is the last point of the SID, being the 7th revision of the procedure and A corresponding to a certain departure runway, for this case runway 30).

The definition of a SID includes both lateral and vertical navigation, imposing required vertical restrictions to ensure separation with terrain. It will thus be common to require minimum climb gradients along certain SID phases. For the definition of these climb gradients, the designer will assume that the aircraft is fully operative. Any contingency procedures required to be followed in the event of an emergency must be designed by the aircraft operator (e.g. departure route in case of engine failure during take off, after V_1). This means that contingency procedures for limited aircraft performance cases are not published in the AIP. Instead, they are particular of each aircraft operator. Departure trajectories under an emergency are, from the ATM perspective, fully treated in the tactical phase, increasing the associated uncertainty. ATC do typically not know what departure in case of emergency an aircraft is going to fly, since these procedures are particular of each operator and may also depend on the ability of the aircraft to turn more easily to one side or the other[1].

SIDs are designed considering:

- **Obstacles** - the presence of obstacles will establish restrictions on minimum waypoint overflying altitudes or vertical gradient restrictions (which are named PDGs, for Procedure Design Gradients) to avoid terrain encounters. Pilots can typically monitor and control their climb rate (measured in feet per minute). However, climb rates do not determine the vertical trajectory of the aircraft completely since this will depend on the Ground Speed of the aircraft (which in turn, is a function of the True Air Speed and the spot wind). SID vertical climb restrictions are thus defined through *climb gradients* instead of *climb rates*. Climb gradients are normally determined using percentages, defining the minimum slope of the required vertical profile (see Figure [14.1]). Unless otherwise published, a Procedure Design Gradient (PDG) of 3.3% is assumed. If a gradient greater than 3.3% is required, it must be made clear on the SID chart and also the limiting altitude to which it extends.

- **Available Navaids** - trajectories have to be defined in a way that aircraft can fly them. This includes taking into account the equipment onboard. SIDs will thus be conventional (based on VOR, NDB or DME navigation) or RNAV. Depending on the accessibility to the navaids (both on the ground or in space), certain maneuvers will be available.

- **Airspace restrictions** - the existence of D, P or R areas will drive the lateral and vertical profile of the SID, avoiding these areas. Also, for ATM

---

[1]When under an engine failure situation, pilots will require to turn against the operative engine

or even ATFM reasons (avoid congested ATC sectors) SIDs may have lateral or vertical restrictions. Using different techniques, and combined with restrictions applied to other departure or arrival procedures, one can design maneuvers in such a way that separation between aircraft is procedurally achieved (the restrictions imposed to the procedures themselves guarantee the separation between aircraft).

There are two types of SID:

- **Straight departure** - one in which the initial departure track is within 15° of the alignment of the runway. Track guidance must be acquired within the first 10.8 nm from the departure end of runway (DER).

- **Turning departure** - one in which an initial turn of more than 15° is required. Straight flight is assumed until reaching a height of 400ft. Track guidance must be acquired within the first 5.4 nm after the completion of the initial turns. Figure [14.2] shows an example of a turning departure.

Instants at which turns must be initiated can be defined either when reaching a determined altitude or when overflying a given fix. These are named *turning altitudes* or *turning points*. Turns must be constructed according to ICAO rules, which take into account the uncertainty associated to the existence of winds, different aircraft speeds, pilot reaction and bank establishment times.

Vertical restrictions in a SID must be plotted according to the criteria described on Figure [14.3].

2 Standard Instrument Arrivals

A Standard Instrument Arrival (STAR) is defined by ICAO as *a designated instrument flight rule (IFR) arrival route linking a significant point, normally on an ATS route, with a point from which a published instrument approach procedure can be commenced.* It is thus a procedure that permits the transition from the enroute phase to the approach phase. This means that the first point of the STAR procedure will be the last point of an ATS route, and the last point of the STAR will be the Initial Approach Fix (IAF) of an approach procedure.

STARs are named using the identification of the first waypoint, a version number and a letter indicating for which runway the STAR is intended (e.g. BAN4D is a LEMD arrival procedure when operating under North Configuration, while BAN3B is the equivalent STAR procedure when operating under South Configuration).

Equally to SIDs, STARs are defined based on obstacle clearance, navaid availability and airspace restriction criteria.

Minimum altitudes are assigned to each of the different legs of a STAR procedure. The Minimum Sector Altitude (MSA) provides at least 1000 ft obstacle clearance within 25 nm of a defined point. MSAs are defined for each aerodrome centered on the ARP[2] or a significant ground navaid for the approach. Figure [14.4] illustrates an example of a MSA definition.

3 Instrument Approaches

The design of approach procedures comes determined by the distribution of terrain in the vicinity of an aerodrome, the expected type of operations (VFR or IFR) and aircraft types. When designing an instrument approach procedure, obstacle clearance must be a primary safety consideration.

3.1 Types of approaches

3.1.1 Precision and Non-Precision

- Precision Approach (PA) is one in which vertical guidance is provided. An ILS approach is an example of a PA, where both, lateral (localizer) and vertical (glide path) guidance are provided. GNSS or GBAS based approaches can also be precision approaches.

- Non Precision Approach (NPA) is one in which vertical guidance is not provided. A VOR approach in which only lateral guidance is provided is an example of a NPA. Vertical profiles for NPAs are defined through timing or DME based step descents.

3.1.2 Straight-In and Circling

- *Straight-In* approaches are those which are aligned with the runway center line. For the case of non-precision approaches, a straight-in maneuver is considered acceptable if the angle between the runway center line and the final approach track is not greater than 30º.

- *Circling* approaches are specified for those cases where terrain or other constraints cause the final approach track alignment or the required descent gradient to fall outside the criteria for a straight-in approach. A circling maneuver is a visual flight maneuver that starts when the instrument approach has been completed. Once a circling approach has been initiated, the runway environment should be kept in sight, in order to penetrate the applicable Minimum Descent Altitude (this term will be defined in section 3.3). If after the circling phase of the approach has been initiated, the pilot loses visual reference with the airport environment, he must start a missed approach procedure turning always towards the aerodrome while climbing, until reaching an altitude and position such that he can follow a defined

---

[2]Aerodrome Reference Point

instrument procedure (and that guarantees terrain separation). Due to the orography distribution around an aerodrome, some sectors may be circling-restricted.

3.2 Aircraft categories

Different aircraft need different speed profiles to maintain their flight capabilities. Depending on the speed of the aircraft, different turning protection areas and visibility requirements will apply. Because of this, maneuvers will be designed differently depending on the expected user. Slower aircraft will be able to make turns with shorter turning radius and associated protecting areas.

Aircraft are categorized based on their expected indicated airspeed over the threshold V_{THR} (which is typically equal to the stall speed multiplied by a factor of 1.3). Real values of V_{THR} are different depending on atmosphere conditions, weight of the aircraft, flap configuration, etc. For the only purpose of categorization, we compute this speed for the landing configuration defined by the operator or manufacturer, at the maximum landing weight, sea level conditions and standard ISA. Thus, the category defined for a given aeroplane shall be permanent and independent of changing day-to-day operations.

Aircraft Category	V_{THR} in KIAS
A	$V_{THR} < 91$
B	$91 \leq V_{THR} < 121$
C	$121 \leq V_{THR} < 141$
D	$141 \leq V_{THR} < 166$
E	$166 \leq V_{THR} < 211$
H	Helicopters

Table 14.1: Aircraft categories

3.3 Minima definition

For each individual approach procedure, an obstacle clearance altitude/height[3] (OCA/H) is calculated in the development of the procedure and later published on the instrument approach chart (IAC). The OCA/H is defined differently depending on the type of approach:

- For a Precision Approach it is the lowest altitude or height at which a missed approach must be initiated to ensure compliance with the appropiate obstacle clearance criteria.

---

[3]When defining the Obstacle Clearance Height, we typically refer heights to the elevation of the applicable runway threshold for which the approach is being performed

- For a Non-Precision or a Circling Approach it is the lowest altitude or height below which an aircraft cannot descend without infringing the appropiate obstacle clearance criteria.

By adding a set of operational factors to the OCA/H, one will get, for the case of precision approaches, the Decision Altitude (DA) or Decision Height (DH). For the case of non-precision approaches one will get the Minimum Descent Altitude (MDA) or the Minimum Descent Height (MDH). It must be made clear that a pilot will not continue the approach beyond the applicable operational minima if visual contact with the runway has not been achieved.

3.4 Segments of the approach procedure

Figure [14.5] shows an example of the different approach segments for a NPA.

3.4.1 Initial approach segment

The initial approach segment is defined from the Initial Approach Fix (IAF) to the Intermediate Fix (IF). Track guidance is normally provided along the initial approach segment to allow an interception of the intermediate segment with a maximum angle of 90º for the case of PAs and of 120º for the case of NPAs. The use of "Dead Reckoning" instead of track guidance is only permitted in ILS approaches where an operational advantage is expected, intersecting the localizer at 45º and being the length of the "DR" leg of not more than 10 nm.

During the initial approach segment a minimum separation of 1000 ft with terrain must be provided. If the geometry of the maneuver does not easily allow to comply with the required maximum intersection angles for the subsequent segments of the procedure, one may design reversal procedures to help with the required smoothness for the aircraft´s maneuver. Figure [14.6] shows the different types of reversal and racetrack procedures. Racetrack procedures are used when the available distance in a straight segment is not enough to accommodate the required loss of altitude. The design of reversal or racetrack procedures has to be made taking into account the uncertainty associated to speed restrictions, bank angle establishment times, pilot reaction times, wind effects and maximum allowed descent rates.

3.4.2 Intermediate approach segment

The intermediate approach segment is defined from the Intermediate Fix (IF) to the Final Approach Fix (FAF, for the case of a NPA) or to the Final Approach Point (FAP, for the case of a PA). It is during this segment that the aircraft reduces its speed and prepares its configuration for the final approach. It should be made as shallow as possible to help the deceleration. Minimum obstacle clearance criteria is here reduced to 500 ft.

3.4.3 Final approach segment

The Final approach segment is defined differently depending on whether it is part of a NPA or a PA. This segment may be made to an aerodrome (for a circling visual maneuver) or to a runway (for a straight-in landing). It is during this segment that alignment and descent for landing occurs.

- In *Non Precision Approaches* this segment starts at the Final Approach Fix (FAF) and ends at the Missed Approach Point (MAPt). From the runway threshold, the FAF is typically placed somewhere between 5 and 10 nm. In NPA, the final approach segment should deliver an optimum descent gradient of 3°, providing all other terrain separation considerations are compatible. The vertical profile along this segment, when using a VOR/DME navaid, is depicted using several DME fixes and their associated minimum crossing altitudes. The Minimum Obstacle Clearance (MOC) is 250ft.

- In *Precision Approaches* this segment starts at the Final Approach Point (FAP) and ends at the Decision Altitude (DA) or Height (DH). The FAP is generally placed at heights that range from 1000 to 3000 ft above the runway. Maneuvers shall be made in such a way that lateral track guidance is acquired before intercepting the vertical downwards profile. For the case of an ILS maneuver this means that the aircraft must intercept the localizer before the glide path. To ensure the adherence of the aircraft to the maneuver, providing terrain separation[4], descent on the glide path must never be initiated until the aircraft is within the tracking tolerance of the localizer. Protection areas are defined along the final approach segment in a much narrower way than in a NPA. These protection areas are designed assuming that the pilot will not deviate from the ILS indications more than half the scale of the instrument. A fix or facility defined within this segment (typically through the presence of an Outer Marker or a DME fix) is used to monitor the vertical position of the aircraft at a given instant, double-checking it is all coherent with respect to the ILS received signals.

3.4.4 Missed approach procedure

Because of different reasons, an initiated approach may not end with the aircraft touching the runway. For those cases in which the pilot decides that it is not safe to continue with the approach, an exit to the procedure has to be provided (aircraft climbing again, aborting the approach). This is named Missed Approach Segment and is a continuation of the Final Approach Segment, to be followed by aircraft that execute a so-called Go Around maneuver.

Some of the reasons for which a Go Around maneuver can be initiated are:

- Aircraft experiences technical problems (e.g. landing gear not properly extended)

---

[4]which can be critical for the cases of low visibility

- Required visual reference with terrain is not established when reaching the minima

- The approach is unstable or uncontrollable (e.g. high energy approaches where pilots could not decelerate the aircraft in time, strong crosswind components or windshear, etc)

- Runway is not clear (ATC reasons)

During the missed approach phase pilots have to face different challenges and workload in the cockpit accumulates. Basically, they should focus on changing the thrust configuration (applying full thrust), getting the aircraft to climb again, retracting landing gear and flaps accordingly. For this reason, the Missed Approach Procedure (MAP) should be kept as simple as possible. Pilots will not pay too much attention to the navigation instruments during the first instants of a Go Around.

One and only one Missed Approach Procedure (MAP) shall be defined for each instrument approach. The MAP is designed taking into account the presence of obstacles along the missed approach trajectory, assuming typically a climb gradient of 2.5%. For this reason, a missed approach should not be initiated if lower than the applying minima.

The Missed Approach Procedure contains three different phases (see Figure [14.7]):

1. *Initial Phase* - begins at the end of the final approach segment and finishes when the aircraft starts climbing. No turns are defined.

2. *Intermediate Phase* - starts at the point at which the aircraft is already climbing and extends to the point at which a minimum terrain clearance of 50 meters is achieved and can be maintained. It is normally designed as a straight phase, although track angles may be changed by a maximum of 15⁰.

3. *Final Phase* - starts as soon as the aircraft gained an obstacle vertical clearance of 50 meters and extends up to an IAF for a new approach. Turns may be here prescribed.

4 Holding Procedures

Upon request of air traffic controllers (due to airspace congestion) or pilots (extra time required to perform additional cockpit procedures because of technical issues), aircraft may have to hold in the air. Since they just cannot stop at a given airspace point and altitude, holding maneuvers in which they are allowed to circulate are defined.

Holding procedures are considered to be standard when turns are defined to the right. A hold has two straight and two turning legs. They are defined with respect to a fix, placed at the end of the inbound straight leg. Maximum speeds are defined within the hold to keep turning radius under control. Also, straight outbound and inbound legs are determined based on timing or DME techniques. Minimum safe altitudes are defined for each holding procedure to maintain separation with terrain. Figure [14.8] defines the most important holding parameters.

Aircraft will initially fly towards the fix over which the holding procedure is defined. Depending on the relative trajectory of the aircraft with respect to the published holding procedure there are three different ways to join it (see Figure [14.9]).

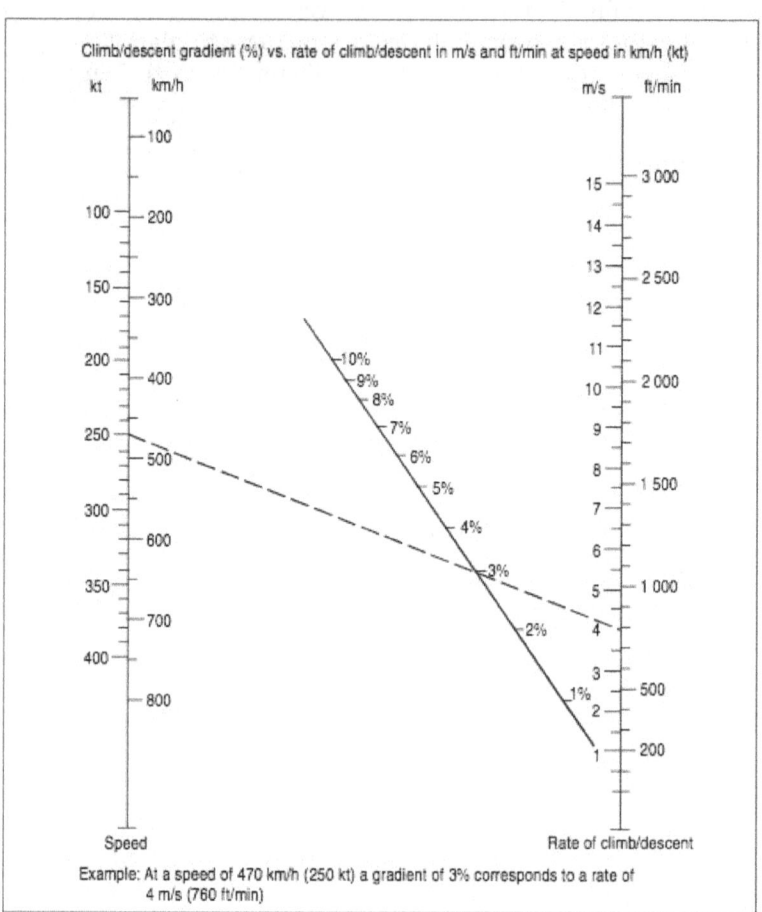

Figure 14.1: Relationship between climb gradients and climb rates depending on existing Ground Speed © ICAO Doc 8168. *Reproduced with the permission of ICAO*

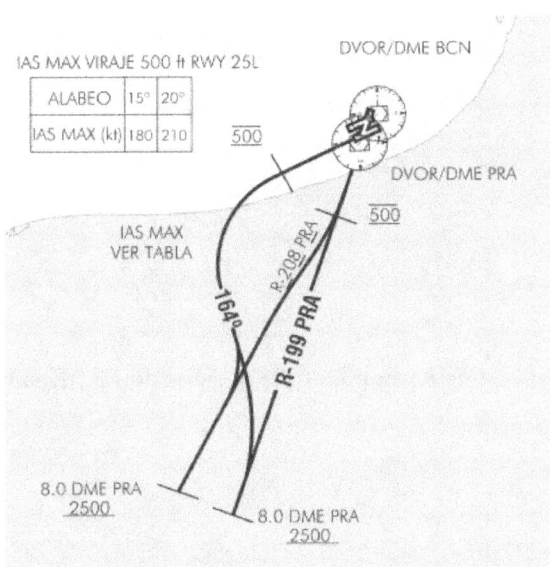

Figure 14.2: Example of a turning departure from LEBL runway 25L. Turn initiated at 500ft AMSL

Altitude/Flight Level "Window"		
	17 000	FL220
	10 000	10 000
"At or Above" Altitude/Flight Level	5 000	FL60
"At or Below" Altitude/Flight Level		
	5 000	FL210
"Mandatory" Altitude/Flight Level		
	3 000	FL50
"Recommended" Procedure Altitude/Flight Level	5 000	FL50
"Expected" Altitude/Flight Level	Expect 5 000	Expect FL50

Figure 14.3: Vertical restriction plotting criteria on SID charts © ICAO Doc 8168. *Reproduced with the permission of ICAO*

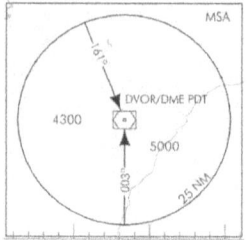

Figure 14.4: Example of MSA definition © Enaire

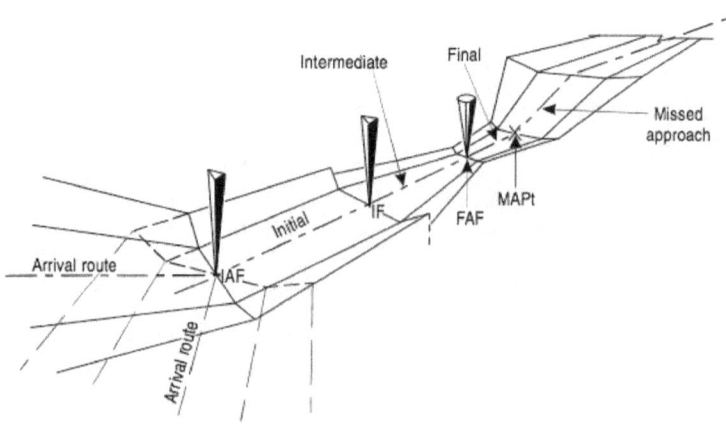

Figure 14.5: Segments of a non-precision approach procedure © ICAO Doc
8168. *Reproduced with the permission of ICAO*

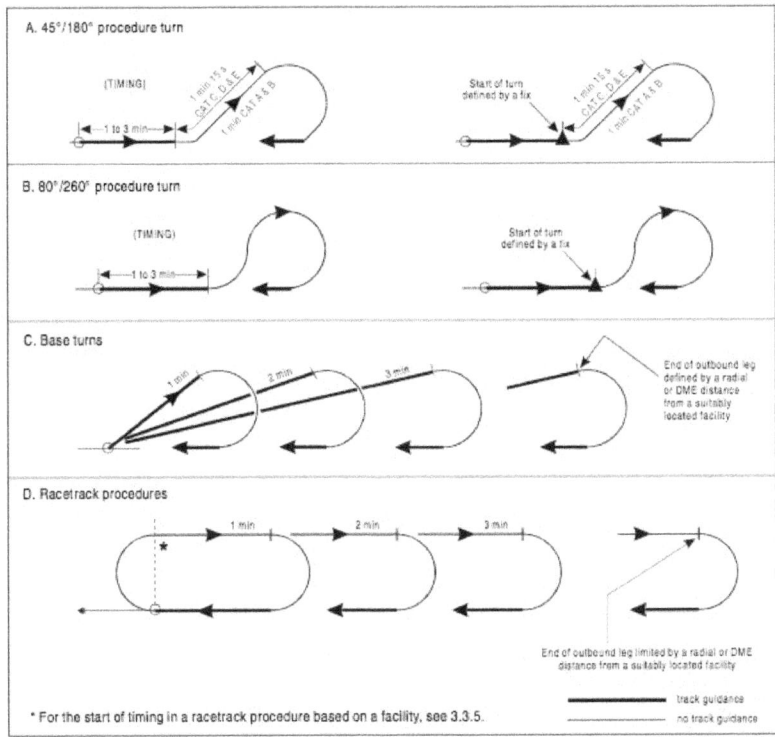

Figure 14.6: Types of reversal and racetrack procedures © ICAO Doc 8168.
Reproduced with the permission of ICAO

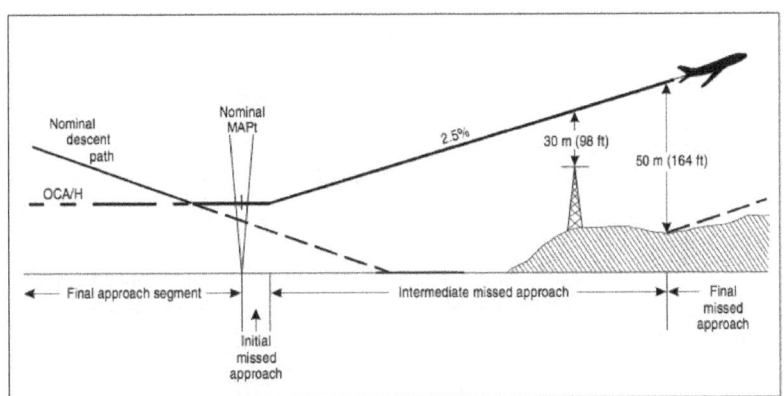

Figure 14.7: Missed approach phases in a NPA © ICAO Doc 8168.
Reproduced with the permission of ICAO

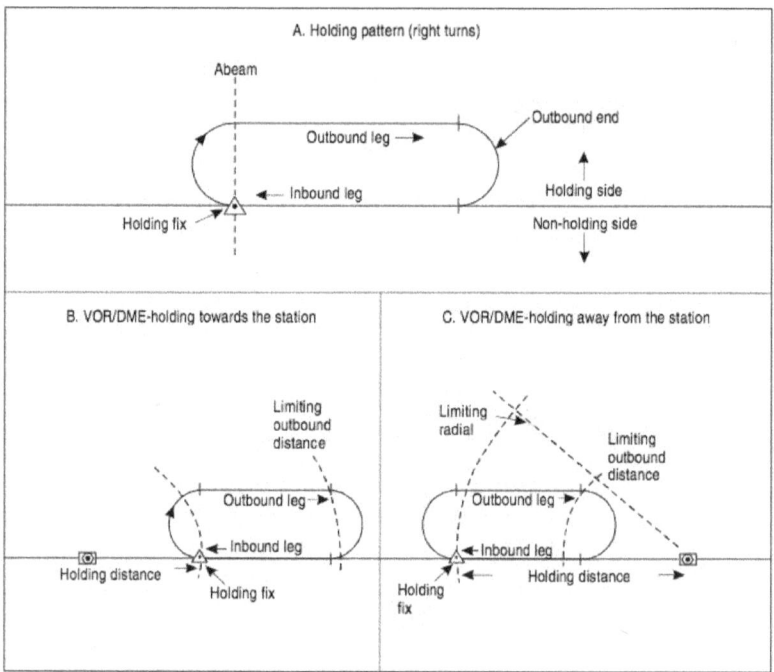

Figure 14.8: Parameters defining a holding procedure © ICAO Doc 8168. *Reproduced with the permission of ICAO*

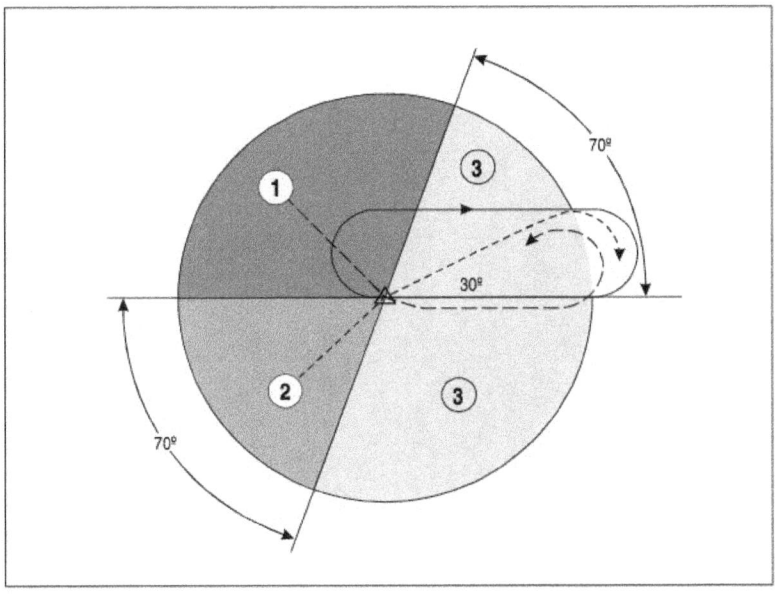

Figure 14.9: Holding procedure entry techniques © ICAO Doc 8168.
Reproduced with the permission of ICAO

Chapter 15. North Atlantic Oceanic Navigation

Although there are many deserted territories on the surface of the Earth, over which the problems that Air Navigation has to deal with are similar, we will here focus on North Atlantic navigation.

The irruption of air transport along the twentieth century meant a completely new way to understand the relationships between Europe and America. We would not have the current figures of trade, relationships and culture exchange without aviation. Globalization would simply not be a reality as we know it today.

Figure 15.1: Charles Lindbergh in 1927 © Wikimedia / Public Domain

The first man who crossed the Atlantic was Charles Lindbergh in 1927. He flew from New York to Paris in a small single-engine aircraft named "Spirit of St Louis". During the 30s, german blimpers surged as competitors against large cruise ships for the small oligarchy that could afford crossing the ocean faster. The Hindenburg disaster (1937), together with the irruption of larger multiengine aircraft (such as the Lockeed Constellation) established a point of no return in the path towards the supremacy of aviation in the North Atlantic. Multiengine piston aircraft did later give way to modern jets such as the Boeing 707 or the DC8. The seventies were domained by aircraft such as the Lockheed Tristar L1011, the DC10 or the Boeing 747. Current ETOPS regulations allow twin engine aircraft to cross the Atlantic (see section 2), enabling the appearance over the Atlantic of the Airbus 330 and 350, the Boeing 757, 767, 777 or 787, among others.

Around half a million aircraft cross the North Atlantic every year, half of which use the OTS (see section 1.1). As we will see in section 1.3, the lack of ground navaids, limited VHF range and the absence of surveillance radars impose new boundary conditions for the development of safe and efficient operations over the North Atlantic.

1 A different airspace structure

As we already saw in our Area Navigation chapter, the structure of an airspace depends, among other factors, on the navigation capabilities of the users and the ability of ANSPs to provide the different ATS services. The lack of ground navaids and PSR/SSR conditions the provision of the different ATS services within the North Atlantic airspace as well as the definition of the ATS routes.

The configuration of the North Atlantic airspace must also take into account two other significant factors:

- **Structure of the demand** - due to the time difference between both sides of the Atlantic and the typical flight times, the demand is structured always in the same way. Aircraft flying eastbound leave the east coast of North America late in the evening (having the passengers the chance to spend their morning and afternoon at their respective american activities), arriving in Europe at the very early hours of the next day. On the other hand, aircraft flying westbound leave Europe in the early afternoon, arriving in America by the early local evening, still having some time to enjoy the rest of the day. This route schedules have proved to maximize the demand and are thus preferred by airspace users (airlines).

- **Presence of the jet stream** - there is a meteorological phenomena that drives the route preference of airspace users. The existence of a constant high speed jet stream over the North Atlantic (blowing from west to east), whose latitude changes every day, makes airlines prefer to fly as close as possible to the jet when flying eastbound (high tailwind components, increasing the Ground Speed of the aircraft, see Figure [15.2]) and avoid it when flying westbound (minimizing headwind impact).

Needless is to say that the structure of the demand, together with the presence of the jet stream at a particular latitude, provokes the existence of peak demand hours over certain regions of the Atlantic, where every aircraft wants to fly the

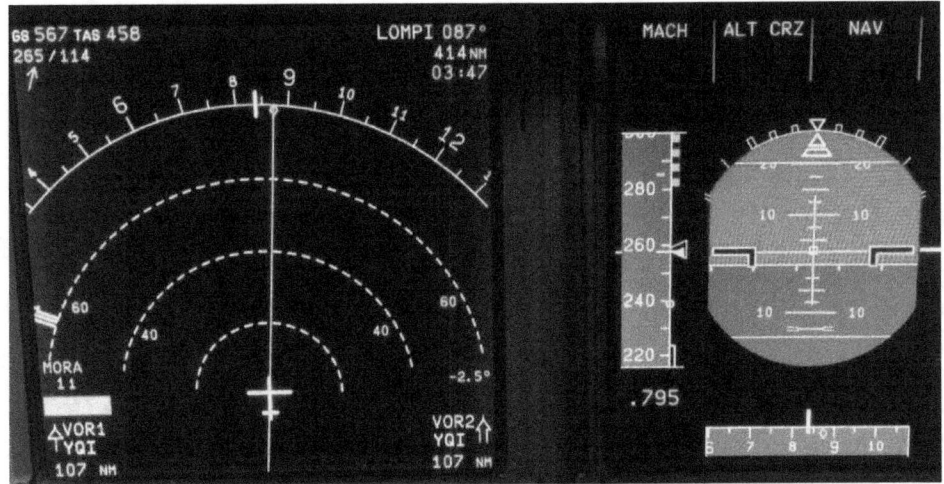

Figure 15.2: A330 aligned with the jet stream flying eastbound (check wind direction and intensity) © Alberto Piquero

same route, at the same level and at the same time. An organization of traffic flows is thus required to keep operations safe.

The jet stream is not always placed at the same place. Its shape, latitude and strength changes on a day to day basis, and so do the intentions of the airspace users. A network of ATS routes is then defined dynamically, depending on the position of the jet stream and willings of the users. This is named the Oceanic Track System and is described in section 1.1. The North Atlantic airspace falls under the responsability of five different Oceanic Area Control Centers (OAC): Reykjavic, Gander, Shanwick, New York and Santa Maria.

These facts are not the same in the South Atlantic region. Time differences and flown distances between Europe and South America are different, changing the structure of the demand with respect to the North Atlantic. Also, there is no jet stream in the South Atlantic. The network of routes defined for the South Atlantic (which is not dynamic, but fixed) has thus nothing to do with the solution found for the North Atlantic.

1.1 The Oceanic Track System (OTS)

In order to accommodate users´ demands with the provision of a safe ATS service, the Oceanic Track System is deployed. A set of separated and organized track structures are published each day for east and westbound flows. These tracks are defined by entry and exit points, together with marked rounded waypoints every ten meridian degrees. Each track is defined also together with its associated available flight levels. The OTS is defined by Shanwick at 2200 UTC of the previous day for westbound traffic crossing the 30W meridian between 1130 and 1900 UTC, and by Gander at 1400 UTC (also of the previous day) for eastbound traffic crossing the 30W meridian between 0100 and 0800 UTC. The existence of oceanic tracks is thus not only dynamic but also intermitent (they are only defined for certain time periods within each day).

The definition of the OTS is done through the NAT Track Message, which includes a Track Message Identification (TMI), that refers to the day of the year for which the tracks are valid (e.g. TMI is 292 for October the 19th). The NAT Track Message specifies the geographic coordinates of the organised tracks, together with the flight levels that are expected to be in use on each track. There may also be details related to the connecting entry or exit routes. The most northerly westbound track, at its point of origin, is designated Track 'A' (Alpha), and the next most northerly track is designated Track 'B' (Bravo), etc. The most southerly eastbound track, at its point of origin, is designated Track 'Z' (Zulu) and the next most southerly track is designated Track 'Y' (Yankee), etc.

```
A ATSIX 62/20 63/30 64/40 64/50 62/60 GRIBS JELCO  EAST LVLS NIL
WEST LVLS 310 320 330 350 360 370  EUR RTS WEST AKIVO
NAR NIL-
B BALIX 61/20 62/30 63/40 63/50 61/60 MIBNO RODBO  EAST LVLS NIL
WEST LVLS 310 320 330 350 360 370  EUR RTS WEST NINEX
NAR NIL -
C PIKIL 56/20 56/30 55/40 53/50 HECKK YAY  EAST LVLS NIL
WEST LVLS 310 320 330 340 350 360 370 380 390  EUR RTS WEST NIL
```

Figure 15.3: Example of westbound OTS definition

The use of OTS tracks is not mandatory. Currently about half of NAT flights use the OTS. Aircraft may fly on random routes which remain clear of the OTS or may fly on any route that joins or leaves an OTS track. There is also nothing to prevent an operator from planning a route which crosses an OTS track. However, although ATCOs make strong efforts to accommodate random traffic across the OTS, significant changes in flight level or reroutes from those planned are typically necessary during most of the OTS traffic periods.

1.2 Minimum Navigation Performance Specifications

Minimum Navigation Performance Specifications (MNPS) are a set of requirements established to fly within the North Atlantic region (NAT). Associated airspace class and RVSM requirements do also apply. Minimum operator, aircraft and crew capabilities are required to ensure the integrity of lateral and vertical navigation performance. These come determined by ICAO Doc 007.

- **Longitudinal Navigation** - within the NAT region, longitudinal separation is achieved making use of procedural air traffic control techniques that control the Actual Time Over (ATO) certain waypoints, together with the

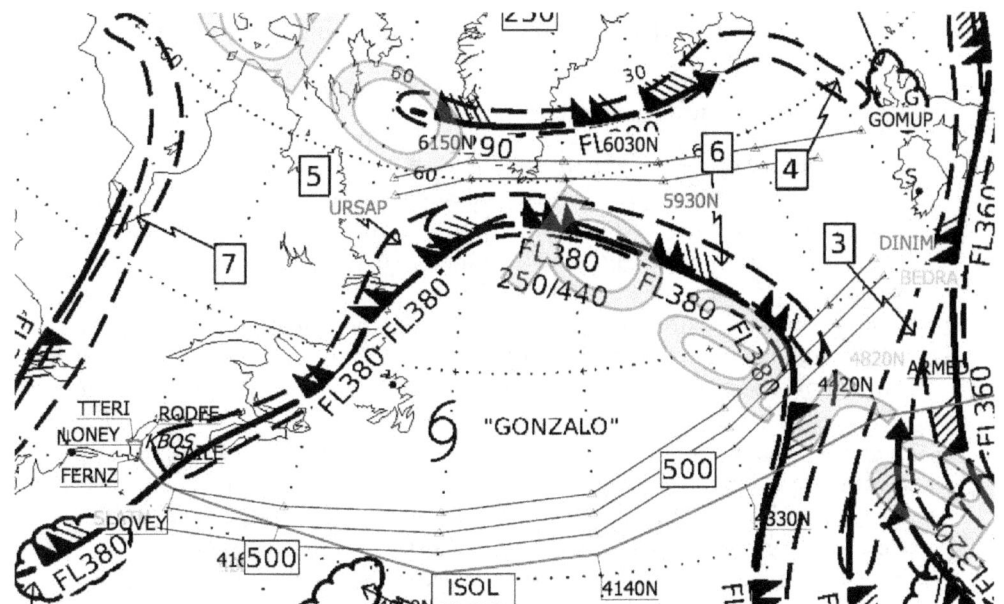

Figure 15.4: Westbound OTS (blue) avoiding the Jet Stream. Flight Plan route not using the OTS System (red)

Mach Number technique (aircraft flying the same route at the same level are required to maintain a Mach Number restriction to avoid overtaking). It is of extreme importance that all aircraft have their clocks synchronized to UTC, working all under the same reference framework. MNPS specifies maximum clock tolerance errors.

- **Lateral Navigation** - onboard navigation equipment that has been certified as RNP 10 or RNP 4 complies with MNPS restrictions. Besides the primary system, it is necessary to have a second (standby) navigation equipment. The required accuracies may be achieved using Long Range Navigation Systems, namely INS, IRS or GNSS.

- **Transponder** - some MNPS regions have radar coverage. For surveillance purposes, it is required that all IFR aircraft operating within the MNPS region have an operating Charlie mode transponder.

- **Communications and Surveillance** - for some specific tracks and levels CPDLC & ADSC are required (see section 1.3).

1.3 CNS concept applied in the North Atlantic context

Continental solutions to the CNS paradigm cannot be applied in the NAT region because of different natural restrictions.

- **Communication** - while continental verbal communications are based on the use of VHF channels (frequencies assigned to each ATC sector), these frequencies suffer important range limitations that make them unusable in larger airspace sectors, such as the ones used in the NAT region. Instead, communications are made using HF channels. These are lower frequency bands, that can more easily bounce against the ionosphere, increasing the range of the signal. The impact of solar radiation on the ionosphere makes some frequencies be more appropiate during the day or the night. A dynamic system of frequencies is established (sectors have a set of HF frequencies that the radio operator can change along the day). HF communication requires constant carrier monitoring and is extremely uncomfortable. For this reason, pilots do only turn on their HF equipment when they have to transmit a message or when they receive a SELCAL alert[1] from the controller, who may want to give them a verbal instruction. Through the use of satellite links a new system was developed, enhancing the quality of the transmission and reducing workload, it is named CPDLC (Controller Pilot Datalink Communication) and is a text system of communication through which reports and clearances are issued (see Figure [15.5]). Similarly to during continental navigation, communications in the NAT region are used to transmit position reports (necessary for procedural surveillance), issue ATC clearances (lateral, vertical and speed control), report weather phenomena or any other abnormal navigation circumstance. In the NAT region, pilots do not talk directly with ATC, instead they speak with radio operators that later relay the message to the corresponding ATCO. These radio operators do not have executive power and thus cannot make decisions. This delays the application of clearances as an operator is set between the two acting agents (pilot and air traffic controller). However, some small regions within the NAT airspace (those which are closer to the coast) use VHF communications. The 121.5 MHz VHF frequency must at all times be monitored by all aircraft, in order to help any other traffic in the vicinity that may need assistance. For other operational pilot-to-pilot communications (weather reports, useful information about the route, etc) the VHF frequency 123.45 MHz is used.

- **Navigation** - absence of ground navaids in the middle of the ocean forces aircraft systems to make use of GNSS for the determination of position. Together with RNAV systems, the navigation function is performed. Before INS, GNSS and SBAS came true, the determination of an aircraft´s position

---

[1]SELCAL or SELective-CALling is a radio system operating typically under HF frequencies that alerts the crew of an aircraft when a ground station wants to communicate

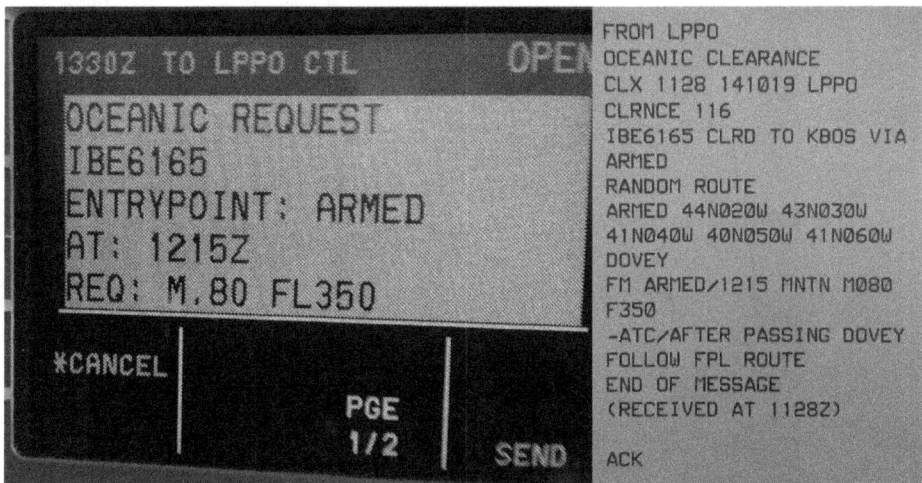

Figure 15.5: CPDLC clearance request by the aircraft (left)
and answer from ATC (right)

when over the ocean was not simple. Systems such as LORAN, OMEGA
or even astronomic observations were used. An accurate determination of
position is essential for a safe development of ATM procedural control. An
example of a position report sent through CPDLD is shown on Figure [15.6].

- **Surveillance** - Most NAT airspace falls out of range for existing PSR/SSR
  and ADSB ground facilities. This lack of radar surveillance forces traditional
  procedural control. However, the existence of satellite links permits the
  use of ADSC (Automated Dependent Surveillance Contract). ADSC is a
  system through which aircraft radiate the position they think they have (it
  is thus a dependent method of surveillance) and use space-based satellites to
  relay these messages to ground stations. These ADSC messages are sent on
  pilot request, ATC request or automatically if the navigation system detects
  an important lack of adherence with respect to the defined route (e.g. in
  case of depressurization, a sudden descent would follow and ADSC would
  automatically send position and altitude reports).

2   ETOPS

During the first stages of aviation piston engines were not as reliable as they are
today. For this reason, regulations imposed that twin engine aircraft could not fly
further than 60 minutes away from a suitable aerodrome. This obviously imposed
important restrictions for oceanic navigation. Twin engine aircraft were forced to

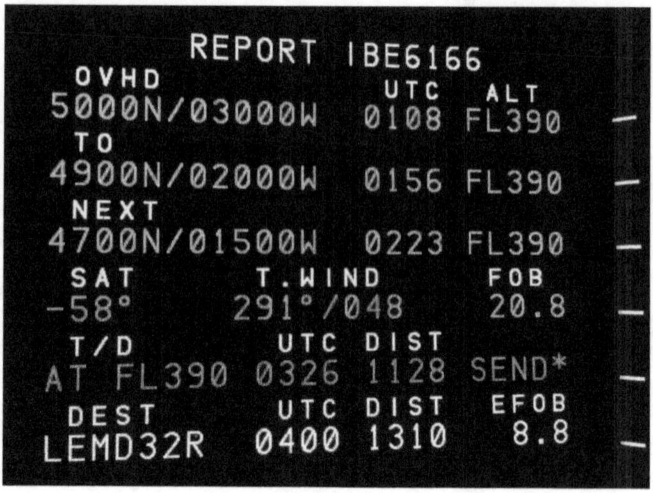

```
          REPORT  IBE6166
  OVHD                 UTC    ALT
  5000N/03000W    0108  FL390    —
  TO
  4900N/02000W    0156  FL390    —
  NEXT
  4700N/01500W    0223  FL390    —
    SAT        T.WIND         FOB
   -58°       291°/048       20.8    —
   T/D        UTC   DIST
  AT FL390  0326  1128  SEND*      —
   DEST       UTC   DIST    EFOB
  LEMD32R   0400  1310   8.8       —
```

Figure 15.6: Position report sent through CPDLC

fly longer routes (through Greenland and Iceland). An efficient oceanic navigation was restricted to three and four engine aircraft. However, turbofans came into being and developed better failure rates, which together with the pressure of the industry towards reducing costs, led to regulators permitting twin engine aircraft to fly further than 60 minutes away from aerodromes, giving birth to ETOPS.

ETOPS stands for Extended range Twin engine aircraft OPerationS. It gathers together a set of requirements to be applied to an airline, aircraft and crew in order to permit the operations of twin engine aircraft further than sixty minutes of flight away from an adequate airport. Depending on the number of minutes a flight can get into a deserted area from a suitable aerodrome, regulators define different ETOPS categories (e.g. ETOPS 120, ETOPS 180, etc).

We would not understand aviation as we know it today if ETOPS had not been developed. Most long range airliners currently overflying the North Atlantic ocean are twin engine ETOPS certified aircraft. With ETOPS 180, 95% of worldwide populated areas can be covered.

2.1 Significant point definitions and critical scenarios

- ETOPS Entry Point (EEP) - first point in the route where the aircraft exits the 60 minute circle centered on the last adequate aerodrome. From this point on, the aircraft would not be able to fly if it were not ETOPS certified.

- ETOPS Exit Point (EXP) - first point, after the ETOPS phase, in which the aircraft penetrates the 60 minute circle centered on the first adequate aerodrome found after the ETOPS leg.

- Equal Time Point (ETP) - Along the ETOPS phase, there is always an airport in mind which is the closest in terms of time to the position of the aircraft. In case of an emergency, the pilot would decide to divert the aircraft towards this aerodrome. The ETP is the point at which the diverting time to two different aerodromes (enroute alternates) is equal. At this point, the enroute alternate for the case of diversion is changed.

- Critical Point (CP) - It is the point, along the enroute phase, in which the required diversion fuel is maximum.

ETOPS regulations require specific fuel uplift policies, in which different scenarios have to be taken into account. The most critical situation foresees the possibility of an engine failure and a depressurization at the same time at any random point of the route. In case this happened, there should be enough fuel onboard to conduct a safe diversion.

Chapter 16. Aircraft Performance influence on Air Navigation

All the concepts we have been reviewing along this text have something in common; something at the core of everything without which Air Navigation would not make sense: the aircraft. Understanding what aircraft can or cannot do will let us design and execute procedures, from the strategic to the tactical phase, that make total sense for the aircraft. A bad understanding of aircraft limitations and crew performance can be at the root of potential latent failures. In this chapter we will focus on how do aircraft operate and why this places restrictions to ATM. We will organize this chapter following the different flight phases.

1 Performance on the ground

When on the ground, aircraft maneuvers are mostly limited because of three main factors:

- Aircraft dimensions - length and wingspan

- Aircraft weight - distribution of weight along the axis of the aircraft determines how these stresses are transmitted to the ground

- ACN vs PCN - Aircraft Classification Number (ACN) comes determined by the aircraft's airport planning manual and specifies (depending on its weight distribution and tyre pressure) the stresses that the aircraft will transmit to the ground. Pavement Classification Number (PCN) determines the capacity of a given ground pavement to support stresses without permanent deformation for ACN values equal or lower than the PCN.

1.1 Apron & taxi

An apron is defined as the part of an airport designed to park aircraft. Aircraft use apron stands for their turn-arounds or sleepovers. These stands are determined by red-painted lines, named aircraft parking envelopes. Stands can be remote or bridged, depending on whether or not the aircraft can be directly connected to the terminal. Low cost carriers tend to use remote stands since they permit the use of two different exit and entry aircraft doors, accelerating the disembarking and boarding processes. Stand size determines the bigger aircraft type that can fit in. Some stands may be incompatible with others, as different geometries and users occupability vary. The assignation of a certain stand to an aircraft is a dynamic and flexible process that has to take into account factors such as the aircraft type, occupability of neighbouring stands, ACN - PCN criteria, etc.

Aircraft departing from remote stands can normally start their engines at the stand and then start moving by their own means. Aircraft placed at bridged stands require ground assistance to perform a *pushback maneuver* (during which engines are normally started). A pushback maneuver is a backwards coordinated maneuver in which the aircraft normally blocks a taxiway while being towed by a special truck (see Figure [16.1]). Turboprop aircraft, however, may be able to perform the backwards maneuver by themselves, exiting the stand using their own engine propellers (reversing the air flow through a proper blade deflection), this is named *powerback maneuver*.

Figure 16.1: KLM pushback track © Barcex / CC BY-SA 3.0

In order to start their engines, aircraft may use their own Auxiliary Power Unit (APU), which provides pneumatic pressure to the compressor of the starting engine, or if unavailable, a Ground Power Unit (GPU) provided by the handling agent. If the APU is operative, it is normally used to start all the engines, but if it were unavailable, the GPU is normally used to start the first engine, and then through a cross-bleeding process, accelerate the already functioning engine in order to bleed air pressure into the compressor of the second engine and then start it up. This is named *cross bleed engine startup* and requires a free area behind the engine that has to be slightly accelerated to supply pneumatic pressure into the other compressor[1].

Figure 16.2: Ground Power Unit © Wikimedia / CC BY-SA 3.0

Once the aircraft has its engines on and gets permission for taxi, the pilot moves the aircraft on the ground using a small steering-wheel that drives the nose wheel, together with differential braking pressure on the main landing gear. Some taxiways might be restricted because of ACN - PCN criteria. Some may have obstacles nearby that limit their use depending on aircraft wingspan or category. Figure [16.3] shows some taxi restrictions as published in the AIP. Aircraft taxi at ground speeds of around 20 knots.

9. TAXIING RESTRICTIONS
- TWY N1 not available for aircraft over 52 m wingspan.
- TWY H1 and H2 simultaneous use is available only for aircraft type C or lower.
- Turning from M1 to S1, and vice versa will be carried out only by aircraft whose key letter is C or lower.

Figure 16.3: Taxi restrictions as published in LEVC AIP © Enaire

1.2 Runway

Aircraft runway performance is critical from the ATM perspective. The times and stages that an aircraft needs to execute prior and during take off strongly affect the runway throughput.

We will first recall the ICAO definition of the three main elements of a departure surface:

- Runway (RWY) - *A defined rectangular area on a land aerodrome prepared for the landing and take off of aircraft.*

---

[1]A simulation of this process can be checked in youtube.com/watch?v=qP_xCNHKhcE

- Stopway (SWY) - *A defined rectangular area on the ground at the end of take off run available prepared as a suitable area in which an aircraft can be stopped in the case of an abandoned take off.*

- Clearway (CWY) - *A defined area beyond the runway, free of obstructions and under the control of airport authorities.*

Each runway has its three RWY, SWY and CWY dimensions defined, with which we can construct the TORA (Take Off Run Available), TODA (Take Off Distance Available, equal to the TORA plus the CWY), the ASDA (Acceleration Stop Distance Available, equal to the TORA plus the SWY) and the LDA (Landing Distance Available, equal to the TORA minus the displacement of the threshold).

Each multiengine aircraft that executes a departure computes the so-defined Take Off Speeds, which are:

- V_1 Decision Speed or Critical Engine Failure Speed - it is the maximum CAS at which a rejected take off must be attempted. Above V_1 take off must be continued unless there is reason to believe that the aircraft will not fly. If take off is rejected after the aircraft has reached V_1, a runway overrun is likely to occur.

- V_R Rotation Speed - it is the CAS at which the rotation of the aircraft should be initiated.

- V_2 Safety Speed - it is the minimum CAS that must be attained when reaching 35 ft height by the end of the TODA.

The computation of these three speeds is not straightforward. Solutions are not unique and an optimization process may be made. These speeds depend on take off weight, runway distance, air density, surface conditions, wind, aircraft configuration, etc. From the ATM perspective it is important to understand that an aircraft will not reject a take off maneuver if it has already reached V_1. Rotation speeds for current airliners (A320, B737, etc) are of the order of 140 knots and take off roll times of the order of 40-45 seconds (from brake release to aircraft rotation).

Since runways may be longer than required, a take off maneuver may be made without the engines running at full power. This is named Flex Take Off or Derated Take Off. It reduces engine wear and operational costs, increasing the engine´s life and reliability. To indicate the amount of required thrust, the pilot uses the Flex Temp concept. The Flex Temp is a temperature value introduced in the FMC and refers to the maximum amount of thrust that the engines would produce if the outside air temperature (OAT) were the selected Flex Temp. The electronic system will adjust the thrust of the engine to provide as much thrust as it would provide if the outside air temperature were the selected Flex Temp. Since engines provide less thrust the higher the outside air temperature (because of EGT restrictions), Flex Temp values will always be higher than the actual OAT. If the required

Flex Temp is lower than the OAT, this would indicate that the engines cannot produce enough thrust to conduct the take off maneuver (a longer runway would be required, or a reduction in take off weight).

Arriving aircraft will overfly the threshold of the runway at a height of 50 ft and a speed according to the regulation of $V_{app} = 1.3\,V_S$. The required landing distance will depend on many factors such as runway condition, aircraft weight, braking capacity, wind components, etc. From the ATM perspective one would like an arriving aircraft to vacate the runway as soon as possible, in order to permit the next runway operation and thus increase runway throughput. A right rapid-exit design is strongly advisable for efficient landing runway occupancy times.

From an ATM perspective, aircraft performance on the runway are measured through different tools that estimate Runway Occupancy Times (ROTs). Instants defined on Figure [16.4] are critical for a proper ROT definition.

These are, for a departing aircraft:

1. Pilot reports aircraft is ready for departure at the holding point

2. ATCO clears the aircraft to line up and wait

3. Aircraft is lined up

4. ATCO clears the aircraft for take off

5. Aircraft starts rolling

6. Aircraft rotates

7. Aircraft overflies the end of the runway

For an arriving aircraft:

1. Aircraft is established on final approach

2. ATCO clears the aircraft to land not later than after the aircraft has overflown the threshold of the runway

3. Aircraft overflies the landing threshold

4. Aircraft touches down on the runway

5. Aircraft starts vacating the runway

6. Aircraft finishes vacating the runway

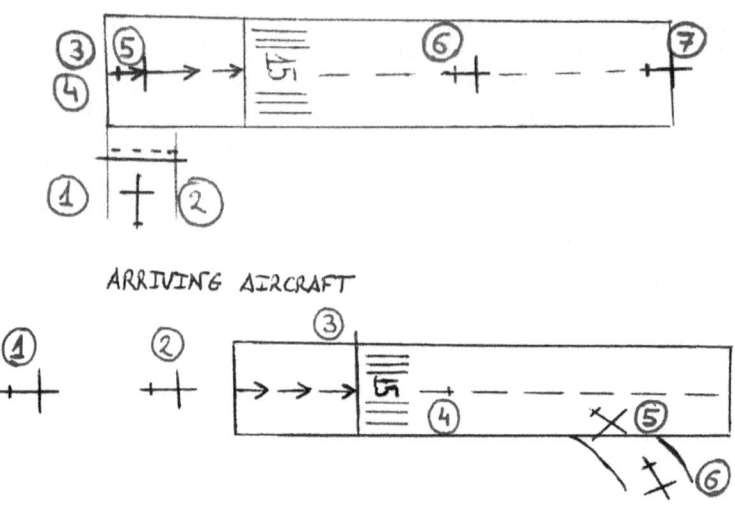

Figure 16.4: Runway Occupancy Time study points

2   Climb & Descent

During part of the climb or descent aircraft follow laterally the assigned SID or STAR. The uncertainty in terms of lateral navigation should be here limited, as we saw in the Area Navigation and IFR Procedure Design chapters. However, even though a minimum vertical profile is required (for vertical separation with terrain), uncertainty associated to the evolution of aircraft's speed and altitude exists. Each aircraft type will evolve in accordance to its own physical limitations, making its behaviour hard to be predicted from an ATM perspective.

Aircraft's weight, atmosphere deviations, aloft winds and cost index, among other factors, will affect the 4D trajectory of an aircraft. Figures [16.5] to [16.10] show how variable the performance of an aircraft can be depending on different conditions[2].

During the descent aircraft would like to perform a Continuous Descent Operation (CDO), which basically consists on applying idle power and let the aircraft glide into the approach. These procedures are published for some airports and may be available during the night, when low demand enables aircraft to perform these maneuvers without interfering in the trajectory of other aircraft.

---

[2]Simulations made for a Boeing 737-800 with a Take Off Weight of 65 tons

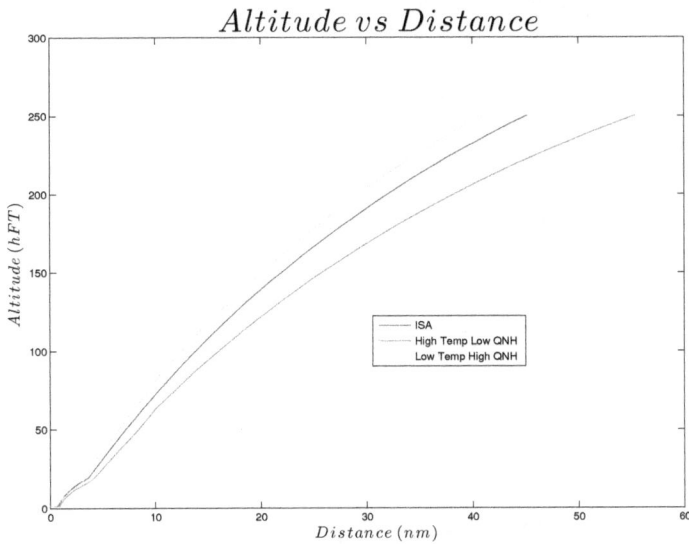

Figure 16.5: Altitude vs Distance climb performance from runway head for three different atmosphere conditions

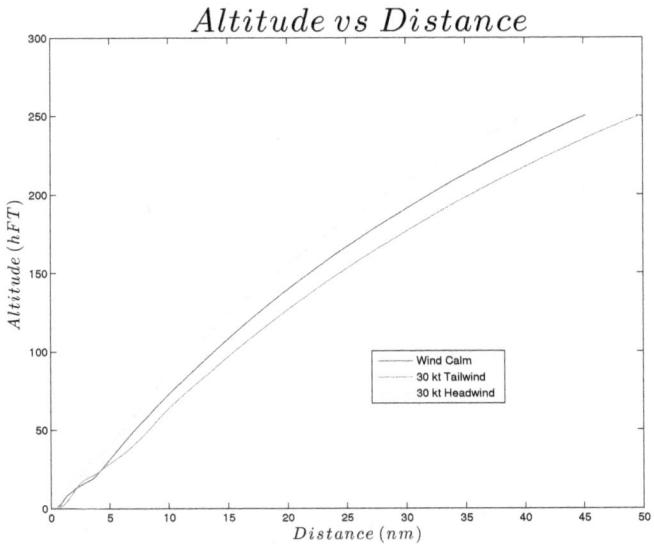

Figure 16.6: Altitude vs Distance climb performance from runway head for three different wind conditions

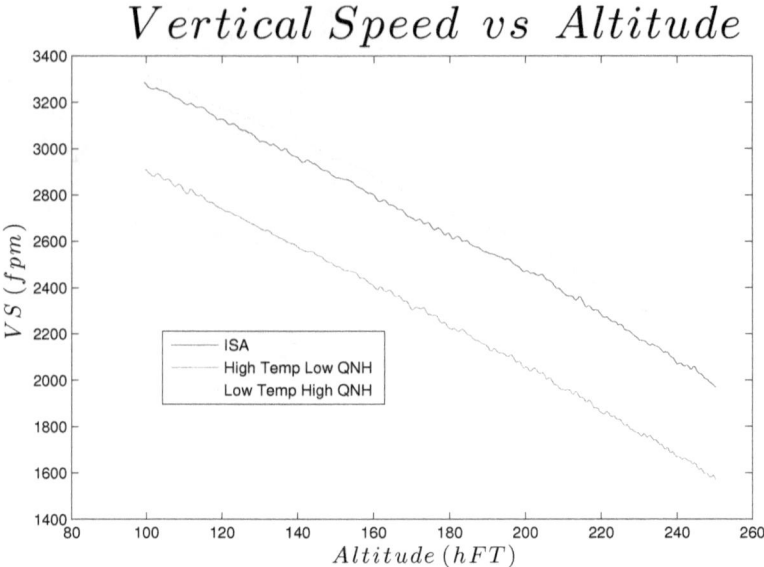

Figure 16.7: Vertical speed evolution during climb for three different atmosphere conditions at maximum thrust and constant IAS

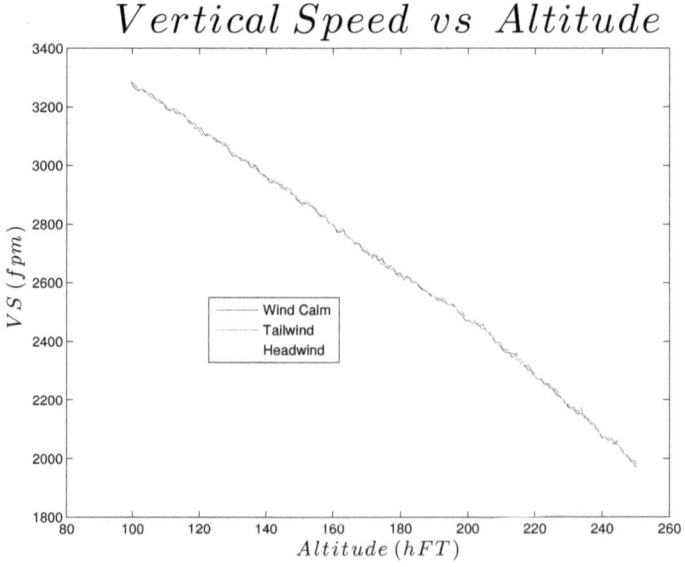

Figure 16.8: Lack of influence of wind on vertical speed evolution during climb

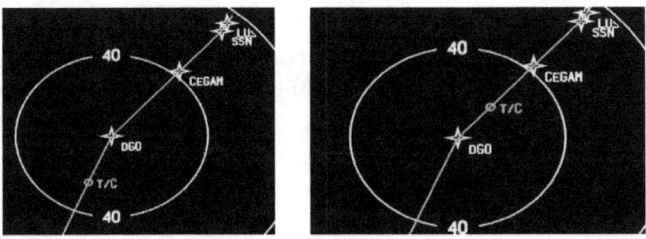

Figure 16.9: Point where the aircraft reaches the cruise level (T/C) for a low (left) and high (right) Cost Index

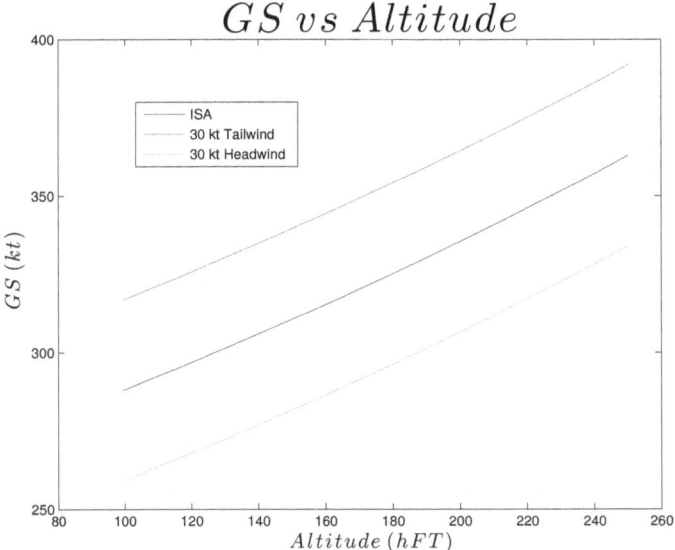

Figure 16.10: GS evolution during climb for different wind conditions under a constant IAS of 250 knots

3 Enroute

During the enroute phase, given the lateral navigation is fixed, the pilot can only try to optimize his flight profile through the selection of a requested Flight Level and Mach Number. The typical objective functions to minimize are flight time and fuel consumption. Trade-off solutions inbetween are found through the use of the Cost Index.

Different aircraft configurations, atmosphere conditions (which determine the available thrust and aerodynamic performance) and gross weight make each aircraft look for a given couple of variables (FL & Mach Number).

Figure [16.11] shows a simulation of the specific fuel consumption of a 737 (fuel kilograms required per nautical mile) for different flight conditions, for a constant Gross Weight of 65 metric tons.

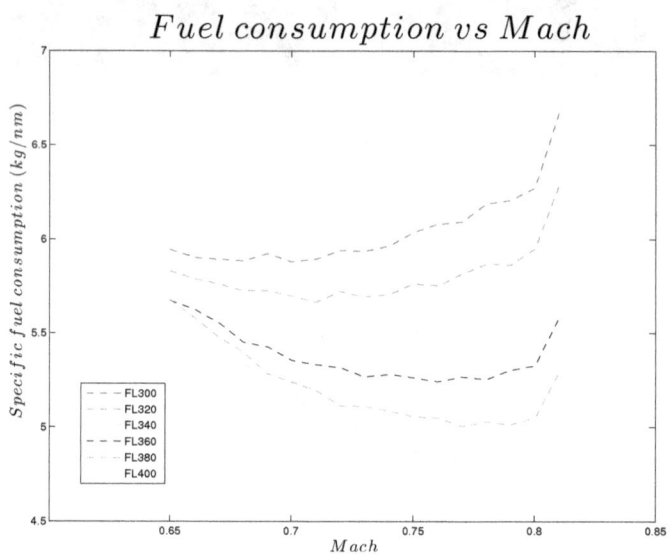

Figure 16.11: Specific fuel consumption as a function of Mach number for different flight levels

For each flight condition and weight, there is an optimum flight level at which specific fuel consumption is minimum. This flight level may not be ATM-appropriate (in a RVSM environment, available flight levels are established every 1000 feet). The required flight level will thus be the closest available level to the optimum.

Since the optimum flight level depends on the weight of the aircraft (increasing as aircraft weight decreases), and this last one is reduced during the flight because of the burned fuel, the optimum flight level increases over time. Pilots request then to make step climbs (typically of 2000 feet each) as soon as lighter weights make it possible. Figure [16.12] shows how specific fuel consumption evolves during a flight[3].

Depending on the established relationship between the costs of time and fuel, aircraft will fly during the cruise phase at a relatively constant Mach number. This Mach number has a lower and an upper limit, that normally come determined by the two possible stalling speeds. This gap of available speeds becomes narrower

---

[3]Simulation made for a B738 climbing at 250 KIAS below FL100, accelerating then to 280 KIAS until reaching Mach 0.78 and then maintaining Mach 0.78 at different cruise Flight Levels

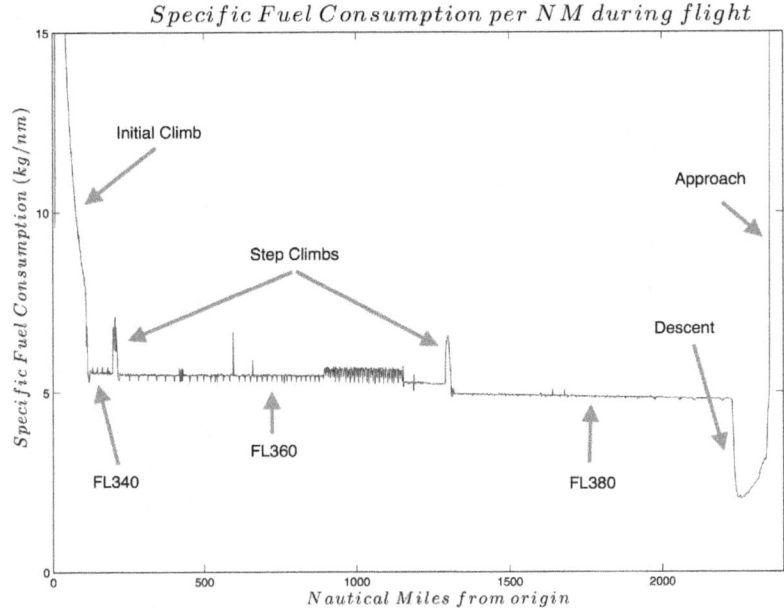

Figure 16.12: Simulation of specific fuel consumption during a flight

for higher levels, and is commonly known as *the coffin corner*. From an ATM perspective we should be aware that for higher altitudes, the Mach numbers an aircraft can make are less flexible.

4 Approach

During the approach it is very important that aircraft stick to the instructed vertical and speed restrictions, for a proper ATM management. It is a critical phase from the ATM point of view since aircraft come closer to each other and have to be sequenced into an arriving runway. During the approach, the crew will configure the aircraft and prepare it for landing (deceleration process, extending flaps and landing gear as required).

If the arriving terminal area were congested, aircraft would have to hold in the air. The only purpose would be then to wait until preceding aircraft land and make room for the next aircraft to approach. While in a holding procedure, aircraft fly at their maximum endurance speed (making fuel flow minimum, in kg per hour). This is achieved at a given performance point, determined to the pilots by an IAS named the maximum endurance speed[4].

---

[4]In the Airbus family, it is commonly known as the Green Dot Speed

It is of extreme importance during the approach to constantly monitor the energy of the aircraft (speed and altitude adherence) since high energy approaches may lead to unstable landings and runway overruns. If at a given decision point, it is considered by the crew that the aircraft has too much energy to continue the approach, a go around maneuver should be initiated.

Different aircraft categories lead to different approach speeds, making the ATC accommodation work harder. The most flexible aircraft in terms of approach performance are the turboprops. Their ability to lose energy lets them fly at almost cruising speeds until very short final, enhancing runway throughput and not requiring too much room behind to accommodate for overtaking faster traffic.

Approach and landing aircraft performance is strongly unpredictable from an ATM perspective. Many decisions are made during this flight phase and many are the factors that affect the variables an aircraft can take. Flap configuration, autobrake selection, reversal thrust braking action, runway slope, runway condition, the existence of windshear, cross or headwind components among others affect the decisions a pilot makes when preparing the aircraft for the approach. They all have an impact in the trajectory and thus on ATM decisions.

5 Efficiency-driven SOPs and impact on performance

In an increasingly competing environment, airlines have developed Standard Operating Procedures that support the decision making of their pilots looking for a more efficient operation from an economic perspective. A proper flight planning process and Cost Index estimation is critical, but also are the tactical decisions that have to be made during the flight.

Here we list some of the procedures airlines are currently introducing their crews into, looking for a more efficient operation.

- Use of ground power equipments when available instead of the APU to reduce fuel consumption while at the gate[5]

- Taxi out and taxi in phases with only half of the engines (switching the rest on just before the departure and switching them off when vacating the runway after landing)

- Rolling take offs in which the aircraft lines up and directly starts rolling (without stopping at the head of the runway, once aligned)

- Pneumatic engine power valves (packs) disconnected during take off to avoid compressor air bleeding and associated loss of thrust

---

[5]An Airbus 340 APU providing full power to the aircraft electric and pneumatic systems uses almost 300 kg of fuel per hour

- Avoid the excessive deployment of flaps and associated drag during take off and landing

Bibliography

1. *Meteorología aplicada a la aviación*. Manuel Ledesma y Gabriel Baleriola. Ed. Paraninfo.

2. *Aerodinámica y actuaciones del avión*. Aníbal Isidoro Carmona. Ed. Paraninfo.

3. *Volar a vela*. Carlos Bravo y Encarnita Novillo. Real Aeroclub de España.

4. *Fundamentals of Aerospace Engineering*. Manuel Soler.

5. *Mecánica del Vuelo*. Miguel Ángel Gómez Tierno, Manuel Pérez Cortés y César Puentes Márquez. Escuela Técnica Superior de Ingenieros Aeronáuticos (UPM).

6. *Spanish Official Aeronautical Information Publication (AIP)*. Enaire. Ministerio de Fomento.

7. *ICAO Annexes 2, 4, 11 & 13*. International Civil Aviation Organization.

8. *ICAO Doc 4444 PANS - ATM*. International Civil Aviation Organization.

9. *ICAO Doc 9971. Manual on collaborative ATFM*. International Civil Aviation Organization.

10. *ICAO Circular 330 on Civil-Militar Cooperation in ATM*. International Civil Aviation Organization.

11. *ICAO Doc 9613. Performance Based Navigation Manual*. International Civil Aviation Organization.

12. *ICAO Doc 8168. Procedures for Air Navigation Services*. International Civil Aviation Organization.

13. *Global Positioning System: Signals, Measurements, and Performance*. P. Misra & P. Enge. Ganga-Jamuna Press.

14. *A-GPS: Assisted GPS, GNSS, and SBAS*. Frank van Diggelen. Artech House.

www.ingramcontent.com/pod-product-compliance
Lightning Source LLC
Chambersburg PA
CBHW071417180526
45170CB00001B/140